T0261170

VIRULENT
ZONES

EXPERIMENTAL FUTURES: TECHNOLOGICAL LIVES,
SCIENTIFIC ARTS, ANTHROPOLOGICAL VOICES

A series edited by Michael M. J. Fischer and Joseph Dumit

VIRULENT ZONES

ANIMAL DISEASE AND
GLOBAL HEALTH AT
CHINA'S PANDEMIC
EPICENTER

LYLE FEARNLEY

Duke University Press · *Durham and London* · 2020

© 2020 Duke University Press
All rights reserved

Printed and bound by CPI Group (UK) Ltd, Croydon, CR0 4YY
Designed by Aimee C. Harrison, Courtney Leigh Richardson,
Drew Sisk, and Matthew Tauch
Typeset in Portrait Text and Trade Gothic LT Std by
Westchester Publishing Services

Library of Congress Cataloging-in-Publication Data
Names: Fearnley, Lyle, author.
Title: Virulent zones : animal disease and global health at China's pandemic
    epicenter / Lyle Fearnley.
Other titles: Experimental futures.
Description: Durham : Duke University Press, 2020. | Series: Experimental futures |
    Includes bibliographical references and index.
Identifiers: LCCN 2019060213 (print) | LCCN 2019060214 (ebook)
ISBN 9781478009993 (hardcover)
ISBN 9781478011057 (paperback)
ISBN 9781478012580 (ebook)
Subjects: LCSH: Influenza—Research—China—Poyang Lake. | Viruses—Research—
    China—Poyang Lake. | Agriculture—Environmental aspects—China—Poyang Lake. |
    Animals as carriers of disease. | Zoonoses.
Classification: LCC RA644.I6 F44 2020 (print) | LCC RA644.I6 (ebook) |
    DDC 614.5/180951222—dc23
LC record available at https://lccn.loc.gov/2019060213
LC ebook record available at https://lccn.loc.gov/2019060214

CONTENTS

ACKNOWLEDGMENTS

Like the scientific research on pandemic influenza that it follows, this book is the product of a journey filled with displacements.

The journey began in New York, where Stephen J. Collier introduced me to the anthropology of the contemporary. I was then fortunate to participate in discussions on biosecurity and preparedness that Stephen organized with Andrew Lakoff, along with Carlo Caduff and Frédéric Keck, which helped shape the research problems and questions I address here.

During my graduate studies at University of California, Berkeley, Paul Rabinow inspired every aspect of this project, particularly the vision of an anthropology that goes to the field not only to find something but also to make something, and remains open to displacements in unexpected directions. I am always grateful for the conceptual and ethical equipment.

At Berkeley this project developed under the influence of wonderful teachers, including Aihwa Ong, Liu Xin, Dorothy Porter, Vincanne Adams, Lawrence Cohen, and Massimo Mazzotti.

During my fieldwork I incurred incalculable debts to my interlocutors at the FAO, in Beijing, and around Poyang Lake. I am particularly grateful for the displacement of my research plans that followed from my early meetings with Vincent Martin at the FAO Emergency Center.

In the drafts through which this book slowly took shape, many people provided crucial insights and commentary. I am grateful to my fellow graduate students in the Department of Anthropology at Berkeley; my former colleagues at the History and Sociology of Science Cluster, Nanyang Technological University; and my current colleagues in Humanities, Arts, and Social Sciences at Singapore University of Technology and Design (SUTD). Nicholas Bartlett, Leticia Cesarino, Ruth Goldstein, Bruno Reinhardt, Anthony Stavrianakis, Hallam Stevens, Laurence Tessier, and Bharat Venkat read drafts both early and late. Emily Chua gave crucial comments and encouragement at many turning points. Warwick Anderson introduced a good number of edge effects to the manuscript from the borderlands of history and anthropology.

It was truly unexpected good fortune to meet Michael M. J. Fischer in Singapore, as he was hard at work cultivating anthropological STS in Asia. Mike provided detailed commentary on the draft manuscript and incredible support for the project over the past several years. I am delighted to be included within the Experimental Futures series that he edits with Joe Dumit.

At Duke University Press, Ken Wissoker provided a sharp eye for synthesis and impeccable clarity during the process of review and revision. Joshua Tranen gave insightful and timely advice. Tim Stallman created the dynamic maps of the 1957 pandemic and Poyang Lake. The book also greatly benefited from the commentary of two anonymous reviewers, whom I would like to thank here.

I am grateful to the institutions that supported this project with grant funding. During my fieldwork in China, I was supported by a Fulbright-Hays Doctoral Dissertation Research Abroad Grant. During the writing of my dissertation, I was supported by a Chiang Ching-kuo Foundation Doctoral Fellowship. The writing of the book manuscript was supported by an SUTD Start-Up Research Grant.

Portions of chapter 3 were previously published as "After the Livestock Revolution: Free-Grazing Ducks and Influenza Uncertainties in South China," *Medicine Anthropology Theory* 5, no. 3 (2018): 72–98. Portions of chapter 4 were previously published as "Wild Goose Chase: The Displacement of Influenza Research in the Fields of Poyang Lake, China," *Cultural Anthropology* 30, no. 1 (2015): 12–35.

Most important of all, I have been inspired and sustained throughout by the vision, sincerity, and love of my parents, Marie and Neill; my brother, Will; and my three Es: Emily, Eliotte, and Edith.

"Is China Ground Zero for the Next Pandemic?"
—*Smithsonian Magazine*, November 2017

With a striking persistence, scientific publications and mass media reports identify China as the possible source of future pandemics. The trope of origins gives the anticipation of future outbreaks a spatial form: suggesting that the seeds of the next pandemic already exist, perhaps hidden, waiting, somewhere in China.[1] The peculiar temporality of pandemic preparedness—focused on potential catastrophic outbreaks rather than already prevalent illness—is an important theme in critical discussions of global health.[2] But what are the spatial consequences of anticipation? How does pandemic preparedness transform the geography of global health research and intervention? And, in particular, how does preparedness differ in those regions of the world marked as sources of disease, instead of the countries that seek "self-protection" from foreign epidemic threats?[3]

To address these questions, this book provides an anthropological accompaniment to the scientific search for the origins of influenza pandemics in China. Adopting the narrative form of the journey or quest, I follow virologists, veterinarians, and wild-bird trackers into the farms and fields of the hypothetical source of flu pandemics. Yet this quest did not result in simple moments of scientific discovery or a definitive arrival at a point of origin. Instead, I show how China's landscapes of intensive livestock farming and state biopolitics created ecologies of influenza that exceeded global health models and assumptions, forcing scientists to reconsider their objects, their experimental systems, and even their own expertise. The search for origins was constantly pushed to the outside, toward new questions about cause and context; knowledge changed when experiments moved along a vector of displacement.

⋮
....
⋮ The search for the origins of influenza pandemics is closely intertwined with the idea of world health and related plans for governance of disease across a global scale. After World War II, the newly formed World Health Organization (WHO) set up an international network of laboratories to monitor "the appearance and spread of influenza" across the planet. In 1957, tested for the first time by a new influenza strain, the WHO's World Influenza Programme traced the so-called Asian flu pandemic to an origin somewhere in China (see figure I.1). Chinese scientists later confirmed that the pandemic began in China, although because of cold war politics China was excluded from the United Nations—including UN agencies such as the WHO—during this period. Eleven years later, the WHO reported that the first cases of the Hong Kong pandemic were identified in refugees fleeing the Cultural Revolution for the British colonial city, again indexing a source in China. However, the WHO's identification of China as a point of origin for pandemic influenza viruses immediately opened new questions: How did the new virus appear? And why did it originate in China and not elsewhere?

In the early 1980s, Hong Kong University virologist Kennedy Shortridge answered that South China could be a "point of origin of influenza pandemics" because of the distinctive ecosystem created by "age-old" farming practices, animal husbandry systems, and wet-rice-paddy landscapes. Shortridge drew closely on laboratory studies, including his own, that suggested human influenza pandemics begin from an animal reservoir. Lab experiments showed, for example, that new strains of influenza could be artificially created by co-infecting a lab animal with two virus strains derived from distinct animal species (such as birds, pigs, or humans). Inside the animal host the two virus strains exchanged genetic material in a horizontal transfer known as reassortment, creating a wholly new strain. Transposing the laboratory model onto China's landscapes, Shortridge argued that South China's farms and fields provided plentiful opportunities for cross-species infections and, therefore, reassortment events: "The closeness between man and animals could provide an ecosystem for the interaction of their viruses." To capture this new ecological concept of pandemic origins, Shortridge called South China the "influenza epicentre."

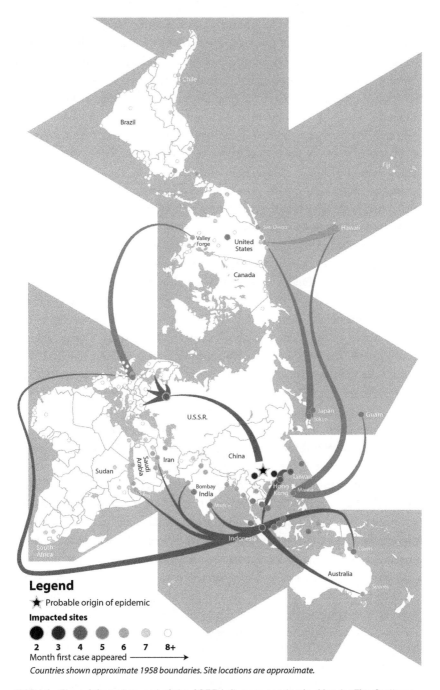

Legend

⭐ Probable origin of epidemic

Impacted sites

● ● ● ● ● ● ○
2　3　4　5　6　7　8+

Month first case appeared ———————➤

Countries shown approximate 1958 boundaries. Site locations are approximate.

MAP I.1. The origin and spread of the 1957 influenza pandemic. Map by Tim Stallman, based on WHO's reconstruction of pandemic origin and spread.

Less than fifteen years later, Shortridge's epicenter hypothesis seemed to be coming true, though not exactly in the way anyone expected. In April 1997, the Hong Kong University laboratory isolated a novel strain of influenza from chickens after disease broke out on three poultry farms in the New Territories, close to the border with the People's Republic of China, sickening and killing thousands of birds. Using tests known as inhibition assays, lab workers identified the hemagglutinin (HA) and neuraminidase (NA) protein subtypes of the strain, enabling them to classify the virus as H5N1. Because H5 viruses are known to be highly pathogenic in domestic poultry, they immediately reported the results to the World Organization for Animal Health (OIE). But H5 viruses had never been isolated from humans or even any mammals, so the lab did not report the finding to the WHO.[4]

Then, in May, a three-year-old boy fell ill with what seemed to be a typical cold. After trips to the local doctor and community hospital did no good, he was admitted to the Queen Elizabeth Hospital in Kowloon. In the hospital's intensive-care unit, his breathing problems increased until, despite mechanical ventilation and antibiotic treatment, he died. Because of the severity of the case, a sample was taken from the boy's throat and sent to the Department of Health (DOH). The DOH identified influenza virus in the sample but could not identify the subtype with existing reagents in its lab. Instead, it sent the virus to the better-equipped WHO reference labs in London and Atlanta, as well as the laboratory of Jan de Jong at the Dutch National Institute of Public Health. Two months later, de Jong called the chief virologist in Hong Kong's Department of Health, Wilina Lim, and told her he was flying to Hong Kong. He had identified the sample as the first known human case of H5 influenza.[5]

As new human cases accumulated, laboratory tests conducted by Shortridge and other researchers showed that the chicken and human viruses were nearly identical. Challenging previous assumptions about interspecies viral transmissions, the virus had apparently jumped directly from birds to humans, probably amid the visceral interspecies exchanges of live-poultry markets. In interviews with the media, Shortridge reiterated his claim that "southern China is the influenza epicentre." News reports began to herald a threatening "bird flu" that could cause the next pandemic. The outbreak was a "pandemic warning," announced a team of virologists that included de Jong and Robert Webster.[6]

On December 27 the Hong Kong government ordered Leslie Sims, assistant director of agriculture quarantine, to kill and destroy all poultry in the territory. Vendors slaughtered chickens at wet markets and left them for government workers to collect in black garbage bags. On farms, government workers gassed entire flocks in their sheds. When I spoke with him a few years later, Sims told me he "probably slept about twelve hours" during that week. He described to me the complex challenges of organizing his staff to go out and kill millions of chickens, especially when many of them, though working in a livestock department, had never even been to a poultry farm before. How to ensure they wore face masks and gloves at all times? Where to dispose of the entire poultry stock of Hong Kong? In total, they killed approximately 1.2 million chickens and 400,000 other birds. After the poultry massacre, no new human cases appeared. The rapid global response seemed to have contained an emerging disease at its point of origin, preventing a potential pandemic.

In fact, the H5N1 virus never disappeared. As Sims explained to me, "Virus continued to circulate in China all through from 1996 to 2004 and the absence of reports of disease does not reflect the true infection status. . . . It is clear from basic biology that disease must have been occurring in the mainland but for whatever reason was not being reported." In 2003 the H5N1 virus reemerged in Hong Kong, in a slightly different molecular form. This time, it quickly spread throughout Southeast Asia, striking Thailand, Vietnam, Cambodia, and Indonesia in quick succession. Later, the virus moved north and west, reaching Egypt and sub-Saharan Africa, Bangladesh, India, and Europe.

As fears of a pandemic grew, a new vision of global health "beyond the human" took shape. In January 2004 the WHO (historically focused on human health), the Food and Agriculture Organization (FAO, historically focused on food security), and the World Organization for Animal Health (OIE, historically focused on animal health) issued a joint statement calling for "broad collaboration" and appealing for international funds in response to the "unprecedented spread of avian influenza." Defining avian influenza as a "serious global threat to human health," the statement explained that if avian influenza "circulates long enough in humans and farm animals, there is increased risk that it may evolve into a pandemic influenza strain which could cause disease worldwide."[7] Within several years the FAO/WHO/OIE would together develop a new strategy for interagency collaboration based on the principle of "One World, One Health," which holds that human health and animal health are "intimately connected," particularly by the zoonotic diseases that spread chains of infection among wildlife, domestic animals, and humans. Proponents of One Health argue that this common vulnerability to disease requires a unified medical, scientific, and governmental response, going beyond the modern disciplinary "silos" of human and veterinary medicine.[8]

Despite the initiation of new interagency lines of communication, however, the WHO continued primarily to fund virus surveillance, vaccine development, and preparedness planning. The FAO, on the other hand, put forward a completely new strategy to control the emerging virus "at source." In an appeal to funders, the FAO declared: "Control of highly pathogenic avian influenza (HPAI) at source means managing transmission of the virus where the disease occurs—in poultry, specifically free range chickens and in wetland dwelling ducks—and curbing HPAI occurrence in . . . Asia before other regions of the globe are affected."[9] In this adaptation of the search for origins, the FAO defined a "source" of pandemics as both a geographic region and a species reservoir, overlaying ecology on geography. With funding secured from donors, the FAO began to build new regional veterinary networks, redeploy staff members, fund scientific studies, and invent new institutional collaborations in Asia and in China. In order to understand and contain the source of emerging influenza viruses, global health moved into the epicenter.

ENTRY POINTS

I began to accompany this movement into the epicenter when I met Vincent Martin, a French livestock veterinarian and career official with the FAO, at the Beijing office of the FAO's newly formed Emergency Center for Transboundary Animal Diseases (ECTAD). Martin established the ECTAD China office (hereafter "the Emergency Center") in 2006 and remained its senior technical director during the main period of my fieldwork (2010–12).

The existence of the Emergency Center and its focus on "transboundary disease" reflected significant internal change within the FAO. "We are a very old, very slow organization," Martin told me, "but the bird flu really forced us to change some things." The FAO had been established after World War II, along with the UN and other agencies such as the WHO, with a special mandate to solve world hunger. Over the years the agency had shifted from provision of food aid toward technical assistance in agricultural development, including some work on disease outbreaks and pest emergencies. Martin himself had many years of experience with control of infectious animal diseases, but the FAO's concern had previously focused on diseases that posed a threat to food security, such as rinderpest or foot-and-mouth disease (FMD). The FAO was not an organization that typically came up in discussions of international or global health, or human pandemics.

After the reemergence of the HPAI H5N1 virus in 2003, all of this began to change. The FAO began to reposition animal health work as a crucial component of pandemic preparedness, a kind of cordon sanitaire at the boundaries of species. "Where animal disease poses a threat to human health," states an

FAO position paper, "FAO's role is to advise on the best methods to contain the disease at the level of animals, prevent its recurrence and undertake research to identify ways of eradicating the disease. . . . The current state of play is that avian influenza is an animal health issue and the focus must be on attacking the problem at source—in animals."[10]

Martin helped to draft the original concept note for the ECTAD system. By collecting information about disease outbreaks from around the world, ECTAD aimed to provide new analytics and advice to both the FAO and member states on emergency response and biosecurity intervention. Initially, ECTAD consisted of a handful of expert analysts based within the Rome headquarters of the FAO, which expanded to hundreds of staff as the avian influenza outbreak spread across the world. However, Martin and others soon ran into a problem: how to validate and interpret the information they collected. As one staff member explained, "We quickly reach the limit of our system. We need expertise in the corridor to recognise what is going on." Some countries offered detailed reports, but others only reported when "everyone already knows."[11]

In 2005 the FAO established the ECTAD Regional Office for Asia Pacific (ECTAD-RAP) in Bangkok. This office became a crucial base for conducting research on avian influenza in southeast Asia, which by that time had spread to Thailand, Cambodia, Laos, Vietnam, and Indonesia. But Martin was not satisfied by this regional presence. He began to travel to China to lobby for establishing an office in Beijing, finally getting approval from China's Ministry of Agriculture in 2008 after one year of meetings with government officials. As he later explained to me, he had been "pushing for having an ECTAD office also in China because I thought that it was meaningless to work in all the surrounding countries, trying to curb the spread of disease, while the epicenter—if we can say so—was in China in a way and it was not good just to have remote collaborations with them, but I thought it was also important to establish an office there." China "was also quite difficult to get in, to have such a close relationship . . . as we had with other countries," he acknowledged. Nevertheless, Martin's arrival in Beijing was the beginning, rather than culmination, of his global health diplomacy.

When I arrived in September 2010, I found the Emergency Center in a sleek high-rise tower on the edge of the Sanlitun diplomatic district, just beyond the East Gate of Beijing's ancient center city. After the formation of the People's Republic of China in 1949, the government had moved foreign embassies from the Legation Quarter, a small *hutong* alley with European-style buildings, to Sanlitun, outside the second ring road. Wide, tree-shaded streets are lined with buildings in a socialist modernist style of gray concrete terrazzo, housing embassies, offices of international organizations, hotels, and residences. Armed military guards in green uniforms are stationed outside each embassy building, adding a subtle hint of contained violence to the peaceful streets. There are also many restaurants catering to expatriate clientele, popular bars and clubs, and tourist-oriented shopping areas such as the so-called Silk Market. Sanlitun is a cosmopolitan space but with a diplomatic cast, reflecting both opportunities of exchange and the sober political negotiations often needed for their enactment. Inside the Emergency Center on the fifteenth floor, the six-person national staff of Chinese veterinarians, statisticians, and program officers worked on desktop computers at cramped cubicles. In Martin's corner office, a large desk with a PC was juxtaposed with a bright red modernist couch. Floor-to-ceiling windows looked out over Old Beijing.

The location of the Emergency Center in Sanlitun reflected the complex international diplomacy that lay behind the movement of global health programs into the epicenter. Looming over everything Martin attempted in China was the recent controversy over China's management of the severe acute respiratory syndrome (SARS) outbreak. In late 2002 an "atypical" pneumonia caused by an unknown virus had spread across southern China's Guangdong Province. But China's government did not inform the World Health Organization of the outbreak until February 2003, after cases of disease had already spread to Vietnam, Hong Kong, and Singapore. Because of the government's continued reluctance to acknowledge the scale of the outbreak, China was widely described as a "global pariah." In response, the WHO announced an unprecedented advisory against travel to affected countries. According to international legal scholar David Fidler, the controversy over reporting led to a "governance revolution" that helped drive the transition from international to global public health. For the WHO and others, SARS demonstrated that control of emerging epidemics should be considered a form of "global public good" that exceeded sovereign state interests.[12]

The SARS crisis also drove a process of administrative and technical reform in China's public health sector. As Fidler has put it, "China was the epicenter of the SARS outbreak; thus, it was the governance epicenter."[13] Once the discrepancy between China's official reports and the actual scale of the epidemic became clear—notably after a whistle-blower, a Beijing military doctor, revealed the number of cases in Beijing hospitals to the international media—China's government reversed course, began cooperating closely with the WHO, built new SARS isolation hospitals, and directed mass campaigns for hygiene and health communication. In the summer of 2003 the outbreak was contained, and China was now considered a "global hero," in part because of the "draconian techniques" used to control the disease.[14] In the aftermath of the outbreak, China realigned public health institutions with international standards of pandemic preparedness, including reconfiguring Mao-era anti-epidemic stations (*fangyizhan*) into Centers for Disease Control and Prevention (CDCs), based on an American model.[15]

Although China's public health sector was increasingly seen as a technically able and cooperative partner with global health agencies, Martin encountered a different set of challenges as he implemented the FAO's plan to control pandemic influenza "at source." Because avian flu primarily infected animals rather than people, epidemic response was largely managed by the Ministry of Agriculture, not Health. "HPAI was just like SARS," Martin complained to me, "but the Ministry of Agriculture hasn't changed." In 2005, for example, the WHO publicly issued a request for timely and comprehensive sharing of virus samples, noting that "from more than 30 reported outbreaks in animals in 2005, no viruses have been made available so far."[16] Much like Indonesia's more famous refusal to share influenza virus samples with the WHO, China continued to assert what Aihwa Ong calls its "national biosovereignty" against global health norms of transparency and sharing.[17]

The movement of experimental systems into the epicenter encountered the legacies of these disputes, leading to the displacement of research toward new forms of scientific communication and collaboration. "Veterinarians don't want to work with medical doctors, and Chinese scientists don't want to share viruses," Martin complained to me in the same breath at our first meeting at the Emergency Center. He described how an initial proposal to sample flu viruses at a lake in southern China was rejected by the ministry, requiring him to work for months to cultivate the right relationships with ministry officials before the proposal was eventually approved. Despite the optimism of the catchphrase "One World, One Health," the world was neither unified nor flat: the geopolitics of territorial sovereignty still governed the pathway to the pandemic epicenter.

The pandemic epicenter carved out a distinctive space where scientific experiments intersected geopolitical territories, reconfiguring knowledge and politics around an exceptional site. On the one hand, the scientific meaning and value of the epicenter were marked as global, because the epicenter was considered the source of pandemics that might spread across the world. On the other hand, the location of the epicenter was inherently singular, a point of origin, and this point was located within China's sovereign territory. The global urgency of pandemic preparedness could be compared to the interventions of humanitarian groups, in which a planetary humanity provides an ethical imperative for constituting "spaces of exception" to the sovereign rule of nation-states.[18] However, the negotiations over access and exchange that I observed led me instead to consider the epicenter as a zone of "differentiated sovereignty," or what Michael Fischer has called a "switching point" between national politics and transnational knowledge circuits,[19] particularly when the objects that Martin and other scientists sought were not as easy to isolate, extract, and transport across borders as influenza viruses.

THE NONVIROLOGICAL

Scientific and popular accounts of the search for the origins of emerging pandemics—including influenza, SARS, HIV/AIDS, and Ebola—are typically narrated as epic tales of heroic virus hunters. In these stories, eccentric and obsessed experts travel to remote and obscure regions of developing nations, particularly sites at the "fringes of the nonhuman world," in order to sample viruses from wild animals, farmed livestock, or the local people. Dressed in full-body hazmat suits, they enter dark bat caves or dense poultry markets, risk bodily contamination, extract viruses, and contain outbreaks. The chaos of the pandemic epicenter, where abominable mixtures give birth to dangerous pathogens, is contrasted with the pure and clear space of the laboratory, where boundaries are preserved, objective knowledge is produced, and danger is controlled.[20]

But when I followed FAO scientists as they moved experimental systems into China in search of the origins of influenza pandemics, what I observed looked nothing like virus hunting. For as scientists got closer and closer to the hypothetical influenza epicenter, the purview of their search expanded in a centrifugal trajectory far beyond the influenza virus to encompass the bodies and behaviors of ducks, traditional techniques of duck husbandry, the geography of rice-paddy landscapes, wild-bird migration flyways, the socioeconomy of live-bird markets, and many other objects inscribed within the ever-widening circles of the ecology of influenza.[21] Rather than traveling to the epicenter in order to bring samples back to the lab, it seemed that scientists felt the need to turn aside and look around, tracing the circumstances and conditions of viral emergence. Their search for the influenza epicenter followed a double movement into China and beyond the scale of the virus, during which research objects shifted from the molecular structure of the virus toward wider zones of virulence.[22]

A few weeks after our first meeting, Martin invited me to a meeting of a United Nations interagency working group, "One Health in China," that he had organized. The meeting took place in the WHO China offices, also in Sanlitun, and included participants from China CDC, China's State Forestry Administration, the Red Cross, and several embassies. In his opening remarks, Martin made an unusual turn of phrase that caught my attention. As he described the global spread of the H5N1 virus from China to Southeast Asia, Africa, and Europe and decried the failure of global institutions to control the outbreak, he pointed to the importance of "nonvirological factors" in the emergence of the H5N1 strain. Migrations of wildlife species, rapid population growth, and an explosion in livestock production, he argued, played crucial roles in the initial appearance and subsequent spread of the new influenza virus. I was struck by this idea of the nonvirological because the term implicitly indexed the predominance of virus-based research in pandemic preparedness. Yet the concept of the nonvirological did not substitute a different causal agent in place of viruses, but outlined a relational approach to viral agency, a virology of the in-between.[23] This concept directs scientific inquiry and global health intervention toward the specific environments of the influenza virus or, put another way, the viral habitat. Instead of studying the virus in the experimentally constructed milieu of the laboratory, Martin highlighted the importance of understanding the actual living environment of the virus in order to understand how, why, when, and where new diseases emerge.[24]

But where could this viral habitat be observed? How could the context of viral emergence be made into a scientific object? At our first meeting in his office, Martin had briefly mentioned the complex negotiations he had undertaken in order to conduct a field research trip at a place called Poyang Lake. I had never heard of the lake before, and I badly misspelled the name in my fieldnotes. Now, in his talk, he referred to Poyang Lake again, this time as an example of the nonvirological factors driving the emergence of influenza viruses. China's largest freshwater lake, Martin explained, is both an overwintering site for hundreds of thousands of migratory birds and a large-scale duck-producing region. With a bucolic photograph of white cranes landing near a duck farm projected behind him, Martin argued that the extensive interface between wild and domestic birds at Poyang Lake could promote the transmission of avian influenza viruses across species and therefore drive the emergence of new, more virulent strains. I soon realized that this was not the last I would hear of the birds at the lake. Over the next few months, almost everyone I met who was working on pandemic influenza in China mentioned Poyang Lake. "Poyang Lake is a perfect storm," warned Scott Newman, a wildlife biologist specializing in the health of migratory birds.

Martin later told me that just as he had begun working to establish the Emergency Center in China, he had read an article written by "Chinese scientists" that brought his attention to Poyang Lake. The main finding of the article reported the establishment of multiple sublineages, or substrains, of the H5N1 virus in southern China. In passing, though, the article also mentioned that the research team had isolated HPAI H5N1 viruses from six "apparently healthy" wild birds at Poyang Lake. Linking the finding with the influenza epicenter hypothesis, the researchers suggested that the birds could be long-distance vectors transporting viruses out of China. And if wild birds were vectors radiating new influenza viruses out of China, Martin knew, Poyang Lake could be a pandemic epicenter.

In the spring of 2006, Martin visited China to make the case for establishing an Emergency Center, and in his presentation he "talked a lot about the Poyang Lake and the potential interest we had in conducting research." By 2010, the Emergency Center already supported a broad range of research initiatives at Poyang Lake, including viral sampling, wild bird tracking, poultry surveys, free-grazing duck movement studies, and satellite image analysis of land use. I was especially interested to hear that many of the scientists traveled to Poyang Lake to conduct these studies. At the lake they captured and tagged wild birds with satellite transponders, counted chicken farms, and measured rice fields. They spoke of Poyang Lake as a fully developed experimental field or, as one ecological modeler put it, as a "geographical unit where we have a critical mass of data to address a question in a new way."[25] The pandemic epicenter was no longer the distant object or objective of a search for the origins of pandemics. It was also becoming the site and venue where that search was conducted. When the chance came, I went, too.

EMBANKMENTS AND INTERFACES

The twenty-seat bus bounced over a high levee and dipped sharply down, following a rough dirt road across bright green wetlands before bounding up another embankment and into Wucheng, a small town on a island in Poyang Lake. Spilling below the south bank of the Yangtze River about halfway between the Three Gorges Dam and the sea at Shanghai, Poyang Lake is China's largest freshwater lake (see figure I.2). Or at least it is during the rainy season. In the wet, summer months, when the lake's vast catchment area swells with rain and the Yangtze rises, the high river pushes water back into the Poyang basin, sometimes causing dangerous floods. But in the winter, when the Yangtze drops, water in the lake ebbs away, exposing vast grasslands in its wake.

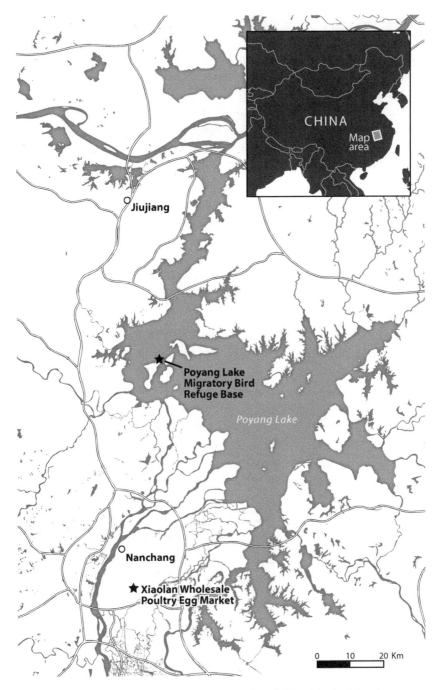

CHINA

Map area

Jiujiang

★ Poyang Lake
Migratory Bird
Refuge Base

Poyang Lake

○ Nanchang

★ Xiaolan Wholesale
Poultry Egg Market

0   10   20 Km

MAP I.2. Map of Poyang Lake. I conducted most of my fieldwork in the intensive rice- and duck-farming counties between Nanchang and the lake. Map by Tim Stallman.

Through these complex hydraulics, the depth of the lake fluctuates as much as fifteen meters, and the surface area covered by water during the flood doubles that of the dry season. The dirt road that my bus followed to cross the wetlands, built several years earlier, is passable only during the dry season. In the wet season, Wucheng is an island.

The lake's peculiar expansion and retraction support a distinctive ecosystem. When the water retreats in late autumn, an enormous green meadow slowly emerges, filled with the exposed roots of water plants and the young shoots of wetland grasses. By November, these green fields attract the eyes of migratory birds flying south from breeding grounds in Mongolia and Siberia. According to estimates, more than 350,000 birds from 105 species overwinter at the lake, including the critically endangered snow crane (*Grus leucogeranus*).[26] Since ancient times, the lake has been a famous site for poetry and landscape painting, often featuring images of soaring wild birds and rising mist. More recently, government decree designated a section of the lake as one of China's first wild-bird refuges and placed the refuge headquarters—including offices, a museum, and a hotel—in Wucheng.

But the lake region is also a "working landscape," a place where centuries of land reclamation and irrigation works have rerouted flows of water and farming systems have transformed ecological communities of plants and animals.[27] Some of the earliest archaeological evidence of rice cultivation in the world comes from sites near the lake, and integrated rice–duck farming dates back centuries. In the 1950s and 1960s, rural residents built enormous embankments during mass mobilization campaigns, reclaiming agricultural land and constructing new irrigation networks.[28] More recently, Poyang Lake has also been caught up in China's "livestock revolution," a term introduced by FAO analysts to describe the growth and intensification of animal production across the developing world. Much like the earlier Green Revolution, modern strategies of technology transfer—including hybrid breeds, manufactured animal feeds, and pharmaceuticals—have begun to disembed livestock farming from environmental constraints, driving intensification of production and enormous growth in outputs.[29]

In both quantitative and qualitative terms, China's livestock sector is perhaps the most dramatic instance of revolutionary change, in part because China's livestock revolution coincided with the country's shift from a planned to a market economy. After the Communist Revolution in 1949, the government organized rural households into production brigades and communes, and smallholder market farming more or less disappeared. Along with rice fields, the commune took over the raising of draft animals and livestock, including pigs and poultry.[30] During political campaigns such as the Cultural Revolution, the state even "forced through reductions in the size of private plots [and] implemented very strict limits on the number of ducks and chicken farmers could raise," according to historian Jonathan Unger.[31] But in 1978, China's political leadership outlined a policy of "reform and opening up" to the planned economy. In rural areas, collective farming was ended and the use rights for cultivation of land distributed to individual households. The state also legalized rural markets, which began to supplant the centralized state procurement system. Poultry was among the first rural products opened up for market trade, along with fish.[32] According to FAO statistics, annual production of meat chickens grew from around 600 million in 1970 to almost 10 billion in 2017, while duck production increased almost fifteenfold, from 150 million to 2.25 billion during the same period. China now accounts for roughly three-fourths of ducks produced in the entire world (see figures I.1 and I.2).[33]

The impact was soon felt at Poyang Lake. In the early 1980s, Jiangxi Province designated the lake region as a "production base" for rice and commercial waterfowl. As villages disbanded collectively farmed land and distributed land-use rights to households, many farmers turned to noncrop activities, such as fish or duck raising. From 1978 to 1998, livestock and fish farming grew from around 10 percent to nearly half of agricultural production in the lake region (by value), and this while overall farm production itself increased tenfold.[34] According to recent data collected from agricultural yearbooks, there are more than fourteen million ducks raised around the lake today—almost half as many as in the entire United States.[35]

When the bus pulled into Wucheng, I walked out of town and along one of the many roads that run atop the embankments. I marveled at these twisting earthwork lines that separate wetlands from gridded rice fields, wild from domestic space. Yet I knew that for influenza scientists, these peaceful embankments could also be understood as dangerous interfaces where wild and domestic birds interact and viruses spread. As Diann Prosser, a wildlife scientist from the U.S. Geological Survey and an FAO research collaborator, has described it, Poyang Lake is a "mixing bowl of people and wildlife and birds." When a virus is transmitted from wild to domestic birds, as Prosser explains, it can reassort or mutate and gain virulence; if the new virus is transmitted back to wild birds, they may "carry it thousands of miles away," seeding a global pandemic.[36]

At the end of the road I found the gated entrance to the Poyang Lake Migratory Bird Refuge. On my first visit to the refuge a few months earlier, Yu, the wiry and gregarious army man who both ran the hotel and led inspections against poaching, insisted on bringing me to see the stuffed rare-bird museum and then

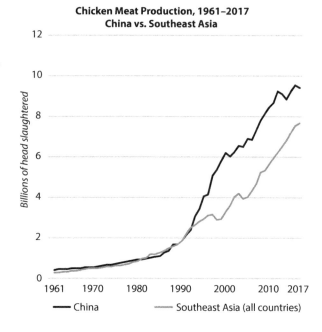

FIGURE I.1. China's livestock revolution in poultry. Graph by Tim Stallman

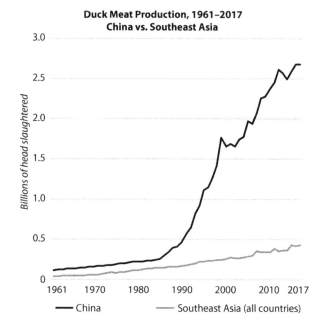

FIGURE I.2. Duck-meat production grew even faster. Graph by Tim Stallman.

out to the refuge to catch a last sight of the migratory birds before they flew north. This time, however, I told him I wanted to see the birds on the other side of the embankment: Poyang Lake's duck farms. The next day we woke up early and, after a quick breakfast in town, Yu drove the white refuge van out along the road back to the mainland, crossing the low passage across the lake. Reaching the end of the lake-bed crossing, Yu drove up the embankment and took a hard right so that we were now driving along the embankment. Pointing out of the driver's-side window, Yu explained that the refuge's core zone was on our right, where only scientific and touristic activities are allowed, and wetlands stretched as far as the eye could see. On the left of the embankment a checkerboard of rice fields and small homes indicated a nearby village within the Experimental Zone, where some economic and production activities could take place.

As we followed the embankment around, a flock of white farmed ducks appeared in a patch of water to our right, swimming inside the boundary of the protected core zone. Immediately ahead of us, I saw two small tents perched on the embankment, and Yu stopped the van so we could get out. Inside one of the tents, on a small raised platform, sat Tang, who greeted us and joined us in the shade of the van to chat. Tang explained he was not the boss who owns these ducks, but a hired technician. The boss, Tang said, had bought about five thousand ducklings, but something like five hundred or more had died from disease. Lacking experience and knowledge of duck diseases, he had hired Tang to take care of the birds. On that particular day, Tang had sent the boss into Nanchang, the provincial capital, to buy medicines.

Tang told me he had raised ducks for over thirty years, first in his home province of Anhui and then in Jiangxi, where he moved when he was twenty-two. I asked him about what had changed about duck raising over the past three decades. He said the biggest change is that back then there weren't so many diseases. Nowadays you really have to raise the birds well, or else they will get sick and die. The problem is pollution. The challenge is that there's no space for duck raising.

Indeed, throughout the Poyang Lake region, the density of duck raising is remarkable. In some villages, nearly every household has a duck shed. As one farmer explained, raising ducks in this lowland area is entirely traditional: "My father raised ducks; my grandfather raised ducks." Yet since the 1980s the scale had changed, he added: rather than ten or at most one hundred birds, today each household raises one thousand, two thousand, or as many as ten thousand ducks. Moreover, the sheds are often clustered around a common water body or along a roadside canal, together creating an even greater scale and density (see figure I.3).

FIGURE I.3. Duck sheds near pond.

China's livestock revolution created enormous growth in animal production but also brought new risks.[37] On the one hand, the increasing quantity and density of animal populations created new opportunities for disease emergence and transmission, leading to widespread outbreaks of porcine reproductive and respiratory syndrome, brucellosis in cattle, and both Newcastle disease and influenza in poultry. On the other, market saturation, increasing cost of inputs, and the industrial restructuring of poultry production intensified market-based uncertainties for farmers, contributing to the growing struggles that in China are referred to as the "three rural issues": an interrelated and multifaceted crisis of countryside, agriculture, and peasantry. Afflicted by flock infections and fluctuating markets, some farmers have abandoned duck raising, leaving empty, ruined sheds behind. Others, like Tang and his boss, have sought out open space, less polluted and at or beyond the very margins of legal land use, including the farming of ducks in the wild wetlands of the refuge.

Faced with new uncertainties, farmers are innovating coping strategies, including the use of pharmaceuticals, relocation of farm sites, and the husbandry of new breeds. As a result, just as scientists moved experimental systems to Poyang Lake and began studying how the "interface" of wild and domestic drove the emergence of new flu viruses, farmers were busily reconfiguring the relationships between wild and domestic birds. The interface, one might say, as much as the disease, was emerging. As scientists reckoned with these unexpected changes to their research objects, epistemological assumptions were put in motion, experimental systems were adjusted, and norms of scientific practice were modified: a process that I call scientific displacement.

SCIENTIFIC DISPLACEMENTS

By describing the movement into the epicenter as a process of displacement, I am extending a spatial metaphor developed in historical and anthropological studies of scientific practice. The concept of displacement has been used to describe the unique trajectory of scientific knowledge production and its transformative effects, and in particular the ways that scientific practice makes new knowledge. As Hans-Jörg Rheinberger explains, experimental systems are characterized by an "economy of epistemic displacement, such that everything intended as a mere substitution or addition within the confines of a system will reconfigure that very system." The production of scientific knowledge is neither a process of discovery, unveiling something that is waiting "out there," nor a process of architectural design in which scientists construct their results according to plan. Rather, scientists create experimental systems that produce "surprises." Experimental systems, Rheinberger argues, produce "unprecedented events" that, although made to happen, also "commit experiments to completely changing the direction of their research objects."[38]

Bruno Latour extends the idea of scientific displacement across a broader anthropological scale. For Latour, *displacement* is a synonym for *metaphor* or *translation*: "the creation of a link that did not exist before and that to some degree modifies two elements or agents."[39] For example, Latour shows how French bacteriologist Louis Pasteur's vaccine for anthrax disease relied on "the displacement of the laboratory" into actual cattle farms, where he conducted field experiments and tests, and then the subsequent "transform[ation] of the farm back into the guise of a laboratory."[40] Successful scientific practice "displaces" or redirects the interests of other actors so that scientific research objectives gain the support of powerful allies.[41] Laboratories create displacements through mastery of scale, as when Pasteur re-creates a cattle farm in miniature inside his lab in order to make microbes visible. In the end the movement of the world through the lab changes the world as well. After Pasteur's demonstration of the microbial cause of disease, for instance, all sorts of problems, from medical practice to urban planning to military strategy, become microbial problems:

"French society . . . has been transformed through the displacements of a few laboratories."[42]

The concept of scientific displacement has been developed primarily based on historical and ethnographic studies of laboratories, prominently including those of Rheinberger and Latour.[43] The choice of laboratories as sites for conducting ethnographic and historical "micro-studies" of scientific practice was in part strategic, because laboratories offered contained spaces where scientific practice and process are readily observable. However, as Karin Knorr-Cetina points out, the laboratory also came to carry a certain "weight" as a key "theoretical notion in our understanding of science." The laboratory came to be understood as "itself an important agent of scientific development" because laboratory displacement enabled observers to explain "the success of science" in terms of everyday practices, rather than theory-driven change or individual discovery.[44] As a result, the model of laboratory practice has subsequently been extended far beyond the "site which houses experiments."[45] In this view, science undertaken in other settings, such as farms and fields, requires the displacement of natural objects onto a controlled or purified site that allows for the "reproduction of favourable laboratory practices."[46] As Latour puts the point most forcefully, "For the world to become knowable, it must become a laboratory."[47]

As I followed the movement of experimental systems into the pandemic epicenter, however, I began to see the contours of another trajectory of scientific displacement that did not begin and end in the laboratory. Julien Cappelle, a graduate student working with the FAO Emergency Center on spatial ecology models, recalled to me his first visit to Poyang Lake:

> The scale was crazy. I mean it was huge, the productivity of the area; the rice, it was, you know, trucks loaded with rice coming from the paddy fields, and in the other way it was trucks loaded with fertilizers and pesticides at a crazy scale. The domestic birds, poultry, it was crazy also; it was like thousands of them every five hundred meters. I've never seen something like that before. What was another thing that was really striking, was, I think it was a farmer that we discussed with, and, because after a while of driving there we saw ducks ducks ducks ducks ducks but no chicken. Which is really surprising because usually you see chicken a lot and ducks less often, and they told us that, uh, there were no more chicken because when you put chicken outside they die. So you're just like, wow, so there's rice everywhere, ducks everywhere, and virus everywhere.

Again and again, when scientists arrived at Poyang Lake, the assumptions underlying their experimental systems were displaced. Scientists quickly recognized that their research objects were also objects of other forms of practice. Farmers were experimenting, too, and at scales that outstripped models or experience. In the laboratory, displacements are "made to happen" by the controlled practices of scientific researchers. But at Poyang Lake, poultry farmers reproduced and changed the objects of scientific research, displacing scientific inquiry into new directions. "Unlike laboratories," historians Henrika Kuklick and Robert Kohler note, "natural sites can never be exclusively scientific domains."[48] The "tracks" that scientists follow in the field are laid in advance, as Cori Hayden puts it, well-worn and "ever-deepening," by the historical and cultural legacies embedded in the landscape.[49] Latour has written that in order to produce scientific facts, researchers should extend their laboratory systems to farms and fields, and make sure not to abandon the protocols of laboratory practice: "never go outside" is the mantra. But at Poyang Lake, it was only when scientists left the lab, looked around, and listened to farmers that they were truly surprised.

At stake is not only a different trajectory of scientific change but also a different account of the scientific subject and scientific agency. Displacement is simultaneously a spatial and a social process. There is an intimate connection between the trajectory of laboratory displacement—a trajectory that Knorr-Cetina specifies as "the detachment of objects from a natural environment and their installation in a new phenomenal field defined by social agents"—and the detachment of scientific experts from broader society.[50] In Latour's account of Pasteur's bacteriology, for instance, the displacement of the microbe from the farm to the lab produced a twofold "reversal of strength": "The change of scale makes possible a reversal of the actors' strengths; 'outside' animals, farmers and veterinarians were weaker than the invisible anthrax bacillus; inside Pasteur's lab, man becomes stronger than the bacillus, and as a corollary, the scientist in his lab gets the edge over the local, devoted, experienced veterinarian."[51] When laboratories extract objects from the environment, they also construct scientists as objective observers of nature. Amateurs, laypeople, and practitioners like farmers or veterinarians are all subordinated to the authority of laboratory expertise.

By highlighting the movement into the pandemic epicenter, rather than the detachment of objects and their circulation back to the lab, this book charts a distinct trajectory of displacement to both scientific objects and scientific subjects.[52] As scientists turned from the virus to the nonvirological and attempted to turn the context of influenza viruses into a research object, their inquiry was repeatedly displaced by their encounters with the artifacts of human practices. Rather than constituting researchers as "scientific entrepreneur-generals

[who] go about waging war to conquer and discipline new allies,"[53] winning carefully staged battles over farmers or local veterinarians through the mastery of scale, the pathway into the epicenter forced scientists to build new connections to the nonexperts that inhabit the research site. In a partial reversal of Latour's reversal of strength, the experts realized how much they needed to learn from those outside of the lab. Not least from farmers and their devoted veterinarians.

AFTER THE EPICENTER

To guide the reader along this pathway into the epicenter, the book is divided into three parts. Each part highlights one layer or strata of the pandemic epicenter, one dimension of the site that causes displacements to the trajectory of global health intervention.[54] Part I, "Ecology," draws on archival research at the WHO in Geneva and interviews to explain the initial movement of pandemic influenza research into the hypothetical epicenter. After showing how laboratory research on viruses played a crucial role in constructing the influenza epicenter hypothesis, I then trace how the current outbreak led to the epistemological displacement of influenza research from virological to ecological disciplines, and follow its spatial displacement from laboratories in Rome or Atlanta to the farms and fields of places like China's Poyang Lake. Within a virological frame, the search for the origins of pandemics had involved a "condensation" of the pandemic threat into a microscopic pathogen. Much like modern biology's treatment of the gene as a metonym for life itself, "the part became the whole." By contrast, the search for the nonvirological environments that produce pandemic viruses drove an expansion in the scale of research objects as scientists sought to understand the "complex systems" from which the virus had previously been "extracted as one tiny part."[55]

However, as Cappelle's shock upon arrival at Poyang Lake makes clear, these contexts expanded in unexpected directions, drawing researchers to question the social, cultural, and political circumstances that shape the ecology of influenza. In the subsequent two parts I show how the scientific movement into the pandemic epicenter encountered displacements produced by two different layers of social and political circumstances. Part II, "Landscape," focuses on the encounters of scientific models and experimental systems with the historical and cultural practice of farm production. These "working landscapes," I argue, continually make and remake the physical environments and interspecies ecologies of the pandemic epicenter. In this part I draw on fieldwork that I conducted with both FAO-affiliated scientists and poultry farmers in the

Poyang Lake region in order to juxtapose the micro-practices of fact construction with the practical configurations of human–environment interaction. Finally, part III, "Territory," examines the intersection of global health projects with China's investments in "biosovereignty," charting how research into the epicenter intersects with national claims over biological resources, agricultural development programs, and state-led veterinary reform. I explore how FAO officers and researchers negotiated with Chinese counterparts to access research sites, virus samples, and information, and I examine how the FAO's efforts to build epidemic response capacity in China's veterinary sector intersected, in unexpected ways, with ongoing post-Mao transformations in the vocation of state-employed veterinarians.

Of necessity, I adopted something like a "multisited" research approach in order to follow the movement of global health into the epicenter. Although Poyang Lake is a central orienting site for both the influenza researchers and myself, other sites such as the Emergency Center in Beijing were equally important in my fieldwork, not to mention the globally distributed locations—Geneva, Rome, London, Hong Kong—that I explored through archival materials. Furthermore, the pandemic epicenter should not be understood as a "local" site confronted by the "global" interventions of international agencies and plans for pandemic preparedness. Rather, the pandemic epicenter is both global and singular, and is stratified by layers of significance that embody different qualities but also cover different spatial scales: ecosystems, regional landscapes, and political territories. As a result, the book is not a documentation of any site or sites, per se, but an account of a journey in search of the epicenter.[56] I examine how the objects of global knowledge become the situated contexts in which knowledge is produced, leading to the emergence of new forms of scientific ethos, livestock production, and political exchange.[57]

During this passage I explore both sides of the doubled relationship between scientific practice and the pandemic epicenter. On the one hand, I document the scientific practices that have made and remade the hypothesis that China is the epicenter of influenza pandemics. On the other, I explore how the hypothesis of the pandemic epicenter produced a displacement in scientific practices by drawing experimental systems into the epicenter and onto the farms and fields of rural working landscapes. Anthropologists and other scholars have recently suggested that scientific and journalistic identifications of China as a pandemic epicenter map a "geography of blame" in which "traditional ecologies, economies and societies figure as 'natural reservoirs' of deadly viruses."[58] As Arthur Kleinman and colleagues put it, "Global discourses regarding the origin and spread of H5N1 avian influenza all too often consist of allegations of blame and assumptions of cultural shortcoming rather than of serious investigations of the political, cultural and socio-economic realities of the societies that have come to be associated with the virus."[59] In this book I investigate China's pandemic epicenter by tacking between sites (Beijing and Poyang Lake) and points of view (FAO livestock specialist, state-employed official veterinarian, duck farmer, etc.) to document the complex interplay among science, blame, politics, socioeconomics, and culture. However, my primary goal is not to provide a better account of the real contexts of the hypothetical epicenter but to follow how scientists and their experimental systems turned these contexts into objects of inquiry. Ultimately, I suggest, only by following scientists and experimental systems in their search for the pandemic epicenter can we hope to reconstruct the constantly reiterated claims that China is "ground zero" for influenza pandemics. In doing so we may also articulate a different understanding of scientific knowledge and expert authority along the way.

PART I

ECOLOGY

Influenza is the paradigm of a pandemic disease, an original exemplar. Before the nineteenth century, the term *pandemic*, meaning "all people," rarely referred to diseases of any sort. Instead, it most often appeared as a critical description of social mores deemed vulgar and common. As Mark Harrison points out, the modern concept of pandemic disease took shape when both medical and lay observers followed the spread of the so-called Russian flu around the world during the years 1889–91.[1] During the Russian flu, news correspondents telegraphed reports that helped readers—and health officials—to track the spread of influenza outbreaks from the Caucasus and Russia to Western Europe, and then around the world to colonial outposts in Africa, India, Singapore, and Shanghai the following year.[2] "This influenza epidemic," wrote Ditmar Finkler, a German physician who experienced the Russian flu firsthand, "broke forth from the East, and overwhelmed the world in a pandemic such as had never before been seen. The high flood of the pandemic flowed over the whole globe in the space of a few months."[3]

Telegraph reports, part of what Harrison calls an "emerging global public sphere" constituted by "new technologies of communication and transportation," made the global scale of the outbreak apparent, enabling observers to distinguish an influenza pandemic from regional epidemics. As a result, the concept of pandemic disease became configured at a global scale, grounded in the emergence of the planet as an object and dimension of knowledge.[4] The search for the geographic origins of flu pandemics was one perhaps unexpected consequence of this global frame.

How does a disease become visible at the scale of the planet rather than the diseased body or the pathogenic virus? Philosophers have emphasized a change of perspective at the core of modern globalism, a view from above that allows one to "take an arbitrary position in the external space from which to approach the Earth like a visitor from a foreign planet."[5] However, as the use of telegraphic communication and mapmaking during the Russian flu makes clear, the planet became a unified, knowable unit only through the mundane spread of standardized infrastructures for observation, measurement, and communication—what Paul N. Edwards has called "infrastructural globalism."[6] Although historians may speak of the unification of the globe by disease as early as the fifteenth century because of worldwide movements of trade and troops, these pandemics did not become objects of knowledge until infrastructures of rapid communication made the global movement of outbreaks observable several centuries later.[7]

Attending to infrastructural globalism is helpful, as Edwards demonstrates, because it reveals how the forms of planetary knowledge change along with developments in technical infrastructure. Initially, observation of flu pandemics depended on the nineteenth-century extension of colonial outposts, including telegraph lines and shipping routes. By the mid-twentieth century, however, a new form of infrastructural globalism would transform the planetary perspective on pandemics: the microbiological laboratory. Andrew Cunningham has described the birth of bacteriological research at the turn of the twentieth century as a "laboratory revolution" that transformed the clinical identity of disease.[8] But in the case of influenza, laboratory transformations took place not at the clinical scale, where patients are diagnosed with a specific disease, but at the planetary scale of the flu pandemic. The isolation and classification of the influenza virus in the laboratory changed how observers tracked the flu, and the flu pandemic, across the globe.[9]

Although several laboratory researchers claimed to have discovered a bacterium responsible for influenza during or after the 1889–90 Russian flu, the devastating 1918–19 pandemic "revealed the limits of laboratory medicine" when no known agent could be consistently linked with cases of disease. After 1918, influenza even became a key model for rethinking epidemics in terms of host factors and environmental conditions of disease.[10] At the same time, the crisis also led to a renewed search for the causal agent, but one that increasingly turned away from bacteria toward the still poorly known, and at the time microscopically invisible, "filter-passing viruses."[11] The term refers to agents that were so small that they could pass through a Berkefeld water filter designed to strain out bacteria. They could not be seen with light microscopes or studied with bacterial culture methods.[12] If a solution that passed through the filter still caused disease in an animal model, the researcher could presume that a virus, rather than bacteria, caused the infection.[13] During the early 1930s, researchers at the British National Institute of Medical Research finally isolated a filter passer that they described as the "influenza virus," producing "a new classification of 'epidemic influenza' as a virus disease."[14]

However, this new viral identity of influenza changed little about the clinical diagnosis of the disease or its medical treatment, which remained "symptom-based."[15] Instead, the laboratory identification of the influenza virus hastened the transformation of the epidemiological identity of the influenza pandemic. In this chapter I chart three key episodes in the laboratory transformation of the influenza pandemic across the twentieth century. In each case, laboratory research on viruses changed how scientists understood the form, scale, and origins of the global pandemic. As the World Health Organization incorporated the results of virus research into a "worldwide" influenza surveillance system, I show how China came into view as the probable origin of influenza pandemics—both geographical source and ecological reservoir—or what eventually would be called the influenza epicenter.

VARIETIES OF INFLUENZA

The discovery of the influenza virus created a new laboratory research object, one that quickly attracted a large number of investigators. During the 1930s and 1940s, most research on influenza was driven by the hope to produce a vaccine, particularly during World War II. The devastating impact of the 1918 flu pandemic, although perhaps "forgotten" by the wider public, was etched in the memory of military leaders.[16] According to historian John Eyler, "The development and testing of vaccines against influenza was one of the most important sources of information in these decades about the flu virus."[17] The transformation of the epidemiological identity of the flu pandemic directly followed from these vaccine research programs but did so because of serendipity and failure rather than achievement of planned objectives.

In 1941, as biologist George Hirst routinely injected virus into an egg in the process of vaccine production, he made an accidental discovery. When influenza-infected allantoic fluid (the egg "white") was mixed with chick blood contained in the embryo, the blood cells clotted and clumped together, an effect known as hemagglutination.[18] After he was able to repeat the effect in a test tube, Hirst argued that hemagglutination could be an effective diagnostic test for the presence of influenza virus in a sample.[19] Even more powerful, however, was a second, related finding: in later experiments, Hirst showed that when blood serum taken from an organism previously infected with influenza—what was known as antiserum—was added to the mixture of allantoic fluid and chick blood in the test tube, the blood no longer clotted. The antiserum, in other words, inhibited hemagglutination. Such a serological method could provide a useful test for identifying whether patients had been infected with the influenza virus: results could be obtained cheaply and within hours, without requiring the painstaking isolation of viruses. However, Hirst's research also raised some new questions. In subsequent experiments, Hirst found that for any particular influenza virus sample, there were some antisera that did not inhibit the hemagglutination process. In those cases the chick blood still formed into clumps. What explained this variability? And was it significant? Part of the answer would come several years later in the aftermath of a devastating vaccine failure.

In 1943 the U.S. Army Epidemiological Board Commission on Influenza, led by Thomas Francis, conducted a large vaccine field trial involving 12,500 students. With evidence that it was working, the vaccine was used on soldiers in 1945, again with impressive results. In 1947 the commission distributed the vaccine once again for use in both soldiers and civilians, but this time observed a wholly unexpected result: the vaccine provided no protection at all. Moreover, virus strains isolated from sick patients showed no cross-reactivity of antibodies with sera from strains isolated in previous years. Francis and other researchers suggested that the epidemic was caused by a new strain of influenza A, significantly different from previous strains, which they called A-Prime.[20]

After the vaccine failure of 1947, Hirst's hemagglutination inhibition assay took on a new importance. If the influenza virus experienced meaningful antigenic variation, then Hirst's HI assay provided an elegant method for measuring this variation. First, the assay allows a researcher to classify two or more influenza viruses in terms of their antigenic relatedness. Compared with other cross-neutralization studies, the HI assay increased the ease and precision with which viruses could be compared. Put simply, if the antisera blocked the viral effect of clumped blood cells, the test revealed an antigenic relation; if it did not, then the virus being tested was antigenically distinct. In time, the great number of distinct viruses identified would lead to new nomenclature as the classification of viruses according to hemagglutinin standards (e.g., H1, H5) was added to the classification by type A, B, and C.

Second, the HI assay also mapped these distinctions onto the terrain of public health and epidemiology. Because the assay uses immunological processes to classify viruses, the results give researchers an indication of the potential public health threat of a sample. In particular, the inability of any existing antisera to recognize and destroy a virus sample suggests that the sample might evade immune systems and infect a large proportion of the population. Looked at from this point of view, one could say that the HI assay is an assessment not of the class of a virus but of the vulnerability of a population.

The vaccine failure transposed the epidemiological difference between epidemic and pandemic onto the antigenic differences among viruses.[21] If a particular influenza virus resists the destructive effect of existing antisera, the virus could be defined as "new" in relation to the immunological history of the human population. Even before a virus spread very far, its potential to cause a future pandemic could be identified. As Australian animal pathologist William Ian Beveridge pointed out, the definition of a pandemic is thereby transformed. "Although the word 'pandemic' means literally just a widespread epidemic," Beveridge later wrote in 1978, "it has come to have a special connotation when applied to influenza: a world-wide epidemic caused by a new subtype of influenza A virus."[22] Defined in terms of viral novelty, this new conception of pandemic changed the infrastructure of the planetary perspective: after 1947, the pandemic potential of a virus could be observed inside a single laboratory.

ASSAYS OF WORLD HEALTH

After World War II, the newly established World Health Organization (WHO) placed this laboratory definition of pandemics at the foundation of a proposed worldwide influenza-surveillance program. At a July 1947 informal meeting in Copenhagen, microbiologists from fourteen countries and a representative from the WHO agreed that the antigenic variation of influenza created a problem that transcended national borders. In theory, a "new" virus emerging in one part of the world could cause a pandemic that threatened the entire planet. For this reason, the prevention and control of influenza had to be international in scope.[23]

For the WHO, surveillance for pandemic flu was one route toward the technical actualization of a moral vision for "world health."[24] When the Interim Commission of the WHO held its fourth meeting in August 1947 to draft a constitution for the incipient organization, one of its first decisions called for the foundation of a "world influenza centre." At the center of the plan was the goal of vaccine development and the problem of influenza variation. As a note by C. H. Andrewes included in the 1947 minutes explains,

> One might perhaps hope to isolate a strain from the beginning of an epidemic, adapt it to growth in fertile eggs and produce a vaccine in time to be of use before the epidemic is over. In practice, there is not nearly enough time to do this within one country. But if it could be shown that a new—and especially lethal—strain was spreading from country to country, the vaccine might be produced in time to protect countries yet unattacked. The above arguments seem to show that many problems concerning influenza can only be solved by international collaboration, such as could be fostered by the World Health Organization.[25]

The World Influenza Center (WIC) initially borrowed space from the British National Institute for Medical Research and appointed Andrewes as director. In the early years the grandiose name belied the fact that the center consisted of "a couple of laboratory rooms and some animal quarters," as Director of the Veterinary Division M. M. Kaplan put it.[26] The key to the WIC's potential ability to provide warning of the next flu pandemic lay not merely in this humble campus, however, but in the international network of laboratories that would isolate and submit viral samples to London: "The Expert Committee envisaged that, by cooperating with an international network of laboratories, WHO could advise member states regarding control of influenza. WHO could also coordinate surveillance on the appearance and spread of influenza in order to accurately forecast the time and place of influenza epidemics."[27]

In most member states, laboratories were equipped with technical equipment sufficient to make a "serological diagnosis as between Influenza A and B, isolating a virus in eggs and drying it off to send to the centre," while at the centre and perhaps a few regional laboratories studies of antigenic variation would be conducted.[28] As Michael Bresalier has argued, the key to this international network was the extension of laboratory equipment and standard protocols to the member states.[29] Andrewes himself stressed that participating laboratories "would have to agree to use common techniques."[30] The WHO's A. M. Payne agreed that "epidemiological reports can be correctly interpreted only in terms of laboratory studies of the viruses responsible."[31] The United States soon lobbied to have a central reference laboratory of equal status with London in Bethesda, Maryland. By 1953, the two "WHO Influenza centers" stood at the apex of an international network involving fifty-four laboratories in forty-two nations and maintained formal or informal contacts with several additional labs in Eastern Europe and the Soviet Union.[32]

The World Influenza Programme (WIP) therefore superimposed two forms of the "globe" in its surveillance of pandemic flu. On the one hand, at the central reference laboratories in London and Bethesda, the potential planetary scale of an influenza pandemic was now rendered in an antigenic difference between viruses visible at a microbiological scale. In a sense, a single laboratory equipped with the proper reagents could identify a "new" antigenic variant that would likely cause a global pandemic. On the other hand, a worldwide network of laboratories, using standard laboratory protocols to collect samples, tracked the appearance and distribution of influenza viruses across the actual surface of the planet.

These two global perspectives did not simply mirror each other. That is to say, the global coverage of the surveillance network was not only a mechanism for verifying the actual global spread of a new virus variant. Rather, the two modes of planetary observation played against each other to construct a narrative of how pandemics appeared, spread, and disappeared. In an interview, Andrewes described how he imagined the WIP would work: "When we're able to trace these movements [of the virus], and particularly the movements of particular varieties of influenza, we are in a position to warn countries just what the virus is like, when it is likely to get there, and what kind of vaccine they should have ready."[33] Soon the WIP's work of tracing the movement and variation of the virus raised questions that exceeded the frame of vaccine development. The 1957 flu pandemic, the first to take place after the establishment of the World Influenza Programme, shows how the WIP's binocular planetary perspective gave birth to a dramatic new hypothesis: China could be the source of pandemic influenza viruses.

ASIAN INFLUENZA

On May 5, 1957, a WHO representative in Singapore sent a telegram to London reporting a large outbreak of influenza in the tropical city:

EXTENSIVE OUTBREAK INFLUENZA SINGAPORE STOP NOT NOTIFIABLE DEFINITE STATISTICS UNKNOWN STOP STRAINS ISOLATED WILL FORWARD WORLD CENTRE[34]

A team of researchers based at the University of Malaya in Singapore isolated a type A influenza virus but could not provide more detailed analysis based on the level of their equipment. They sent samples to the WHO Influenza Center in London.[35] The London researchers subsequently identified a "new variant of type A." On May 24 the WHO sent a memo to all influenza centers in the network: "The virus isolated in Malaya is evidently a new variant of type A differing in several respects from recently isolated strains. . . . Although this is not certain it appears to be so different from previous strains that existing vaccines would probably not give protection. It is therefore being distributed to vaccine manufacturing firms as well as Influenza Centres."[36]

Thus, by late May, although the virus was still confined to the Western Pacific, the WHO had already identified a "new variant"—soon referred to as the A2 or Asian variant—that it expected might cause a global flu pandemic. In the weeks and months that followed, the WHO continued to collect and collate reports of epidemics taking place around the world.[37] After the outbreak was "notified to the whole world," a WHO *Bulletin* later noted, "health services were on the alert everywhere. It was carefully followed as it spread across the world, and in some countries its arrival was even predicted and awaited."[38] Surveillance of the 1957 pandemic thereby linked identification of a novel virus with observation of that virus as it spread across the planet. As a result, monitoring the pandemic in progress inscribed a heterogeneous planetary geography through which the location of appearance and the temporal spread of a pandemic could be mapped. The WHO named the 1957 pandemic the "Asian Influenza Pandemic" because the World Influenza Program's laboratory surveillance network detected the novel variant in Asia and was then able to follow the virus as it spread through populations elsewhere in the world (see figure I.1).

However, the 1957 pandemic also exposed major gaps in the WIP laboratory network. The most crucial of these gaps appeared in the probable source of the virus itself: China. China had been a founding member of the WHO in 1947, at the time under the rule of the Republic of China (ROC). Indeed, China was given the honor of being the first signatory of the UN Charter. But after the Communist Party took power in Beijing and the nationalist government retreated to Taiwan, the United Nations recognized only the Taiwan-based Republic of China. This policy was also followed by all of the UN agencies, including the WHO and the FAO. The rest of the mainland, ruled by the People's Republic of China, was unable to participate in WHO activities, including the World Influenza Program.[39] Although the exclusion of mainland Communist China from the UN had roots in cold war geopolitics, the situation was actually far more historically particular. The Soviet Union and several Communist-ruled Eastern Bloc countries were members of the UN and worked closely with UN agencies such as the WHO.[40] As A. M. Payne put it, "The 1957 pandemic of influenza is the first that it has been possible to study using modern virological techniques in an almost world-wide network of laboratories which had been organized by the World Health Organization with just such an eventuality in mind. It was almost ironical therefore that the epidemic should originate in an area not covered by the programme."[41]

The irony of Communist China's exclusion from the WHO is even greater than Payne makes it out to be, however, for the WHO missed much more than epidemiological reports of outbreaks or virus samples. Indeed, one could perhaps describe Chinese laboratories at this time as an alternate "world center" of influenza research. Chinese scientists, led by microbiologist Zhu Jiming (Chu Chi-Ming), isolated the influenza A virus and independently identified the strain as a novel variant in March, at least two months prior to the WHO's May 24 announcement. Born in Jiangsu, China, in 1917, Zhu completed a PhD at Cambridge University and was a key research fellow working with C. H. Andrewes at the National Institute of Medical Research's World Influenza Centre between its founding in 1948 and 1950. After the unification of China under the Communist-led People's Republic, Zhu—who had several friends in China's Democratic League, a moderate center-left party that eventually allied with the Communists—decided to return to his home country. He was first assigned to the Central Biological Products Research Institute in Beijing and later to the Changchun Biological Products Research Institute.[42]

In Changchun, Zhu soon had an opportunity to continue his research on influenza. In March 1957 he and his colleagues identified an outbreak of influenza in the city and quickly isolated several viruses. The height of the epidemic arrived in mid-March, and by early April the number of cases was declining. Epidemiological evidence showed that the outbreak was extremely widespread, affecting all age groups more or less evenly, suggesting that "the population lacked immunity to the circulating virus." The team isolated thirty-two viruses from patients and found a remarkable 94 percent of the samples positive for influenza. Then, using the same techniques that he had used at London's WIC, Zhu compared these viruses with other flu viruses isolated across China and the world in previous years. These studies demonstrated that "although the viruses isolated in Changchun were still 'A' type viruses, the viruses were antigenically distinct from all other 'A' viruses included 'swine type' (猪型), 'original A-type' (原甲型), [and] 'A-Prime type' (亚甲型)." In serological studies, Zhu and his lab "could not see any relations whatsoever between the new viruses and previous A type viruses." This was, Zhu argued, the most significant variation seen in the influenza virus since the 1946 A-Prime pandemic.[43]

In December 1957 Zhu submitted an English-language report of this research to a medical journal published in Soviet-bloc Czechoslovakia. This report reiterates that the strains isolated in China between March and April 1957 are "antigenically distinct in the hemagglutination inhibition test, from swine, classical-A and A-prime viruses." Zhu adds that his laboratory was also able to compare these Chinese flu strains with those isolated in Singapore (A/Singapore-1/1957), the Netherlands, and the USSR in May or later and demonstrated that they are "antigenically similar."[44] Zhu concludes that "the 1957 epidemic was caused by a new variant of a virus to which the population possessed little immunity" and must have "originated from one locality and spread along communication lines with human movement." He then raises a fundamental question: "Was it imported or did it rather originate in China?"[45]

Zhu based his answer on the results of routine influenza surveillance ongoing in China in previous years, which he compared with data published by the WHO. During the winter of 1956–57, large outbreaks of influenza were reported across China, and laboratories in Beijing, Loyang (Luoyang), and Changchun isolated the older A-Prime virus from patients. But then, in the spring of 1957, "When the large epidemic swept across China in March–April, all strains isolated and so far studied appeared to be of the new variant type." Drawing from WHO reports, Zhu then tracked outbreaks of the new strain into Hong Kong in mid-April, Taiwan, Singapore, and Southeast Asia, then Australia and South Asia at the end of May, before the new flu virus finally reached Europe and the United States in early June. Zhu concluded that "an epidemic caused by a new variant of influenza A virus apparently commenced in China and spread via Hong Kong to Southeast Asia, Japan, and the rest of the world."

Perhaps equally important, Zhu recognized that identifying China as the source was not the end of the inquiry but rather raised crucial new questions about the origin of influenza pandemics. As he explains, "There is reason to suspect that the new virus actually originated in China, perhaps in the Kweichow [Guizhou]–Yunnan border region where the epidemic apparently started. How and why could such a variant have arisen is a question of obvious importance."[46]

Retrospectively, the WHO acknowledged that the Chinese team identified the novel variant first. In September, a WHO press release relayed information received from Chinese sources about the origins of the epidemic (note the geographical discrepancy from Zhu Jiming's account, however): "It started in the north of China at the beginning of spring and penetrated into the interior of the country, where the virus was isolated for the first time at Peking, in March. Cases next appeared in Hong Kong in the middle of April." As the WHO concluded, "The gaps existing in what should be a world-wide system of epidemiological notification have become painfully apparent on this occasion."[47] As Payne added in his later paper, "It is clear that [the Chinese researchers] recognized most of the important features of the virus which have since been described elsewhere. It is unfortunate that this information did not reach the rest of the world until the epidemic was already spreading widely. If it had we should have had two more months to prepare."[48] From the very first moment when epidemiological and virological research identified China as a possible source of an influenza pandemic, the scientific infrastructures of planetary observation already had become entangled in a different, less than fully transparent, spatial ordering: global geopolitics.

In the next section I show how the spatial identification of China as an origin of pandemic influenza was further transformed and specified by research into the ecological conditions that give rise to new influenza viruses. For, as Zhu Jiming asked, if the influenza pandemic not only commenced in China but the new variant virus also originated in China, what were the conditions of this origin? Much like the planetary perspective, which was made possible by the WIP laboratory network and standard virus assays, this ecology of influenza emerged from the laboratory.

LABORATORY ECOLOGIES

Nowhere is the historical ontology of diseases more uncertain than in the species multiplicity that, we now know, hosts influenza viruses. The history of laboratory research on influenza viruses is for long periods a series of separate histories, each describing research programs devoted to apparently distinct pathogens in different animal hosts. Research on the causal pathogens behind diseases that afflicted humans, swine, poultry, and horses proceeded in parallel, even counterpoint, rather than in unison. As late as January 1933, Walter Fletcher of the British Medical Research Council complained that "no animal (except possibly the anthropoid ape) is affected by influenza," thereby, in his view, reducing the value of experimental research with animal models.[49] During the course of the 1930s, laboratory identification of the swine flu and human flu viruses began to show links between these different diseases and, by implication, ecological connections between these different species. To adopt a phrase from François Delaporte, "an epistemological transformation made it possible to see" these different diseases "in a new light."[50] By the 1950s, American researchers working on swine flu, British researchers studying human flu, German researchers studying fowl plague, and Czech researchers studying equine flu established that these diseases were all caused by similar—even perhaps identical—influenza A viruses. However, researchers remained perplexed by this accumulating evidence: What was the connection between these viruses in different species, if any? Could they transmit from one species to another?

During the 1957 "Asian" pandemic, scientists began for the first time to seriously consider the possible role of animals in the origin of pandemics. As Kaplan and Beveridge later recalled, "Before the 1957 pandemic only a few workers . . . paid any attention to the possible relationship between animal and human influenza. . . . Influenza as a zoonosis was considered a biological curiosity rather than an important link in the epidemiology of the disease in man."[51] Researchers in the People's Republic of China were among the first to draw links between animal flus and the human epidemiology of the pandemic. As Zhu Jiming pointed out, agreement that the 1957 pandemic commenced in China merely displaced the question of origins from "Where?" to "How and why?" When Zhu isolated the new variant influenza A virus, he located the origin of the virus in the southwestern part of China, somewhere in the province of Guangxi or Guizhou. But he also noted that the 1957 "variant" strain was not merely antigenically distinct from the previously circulated strain; it was also an "extremely significant variation."[52] As he later explained, "The new variant virus appears to us to be too radically different in its antigenic and other characters from the A-Prime virus, its immediate predecessor, to make evolution from the latter a plausible hypothesis." Evidence suggested that this virus could not have evolved through the gradual accumulation of mutations during seasonal human outbreaks. Instead, Zhu sought a new ecological explanation for the origin of influenza pandemics. "It would probably be wiser," he suggested, "to look around for some other possible explanation such as an unexpected animal reservoir as the origin of this queer variant."[53]

Soon after the human outbreak, Chinese scientists began looking for influenza viruses in animals as well as humans. In June 1957 reports reached the WHO from Chinese scientists describing the isolation of "the variant strain from the lungs of naturally infected pigs in China, and of epizootics of an influenza-like disease in areas of China severely struck by the human disease." In response to these reports the WHO initiated the first worldwide "animal serum survey" on influenza. As Kaplan later recalled, he developed the survey "because at that time [he] had the strong hunch that animal reservoirs could be very important in the epidemiology of human influenza and origins of mutants or recombinants."[54] In July and August 1957 the WHO requested that veterinary services in a large number of countries "take blood specimens from swine and horses in different parts of their countries, if possible before the pandemic struck, and to take a second specimen from the same animals, if this could be done, about three months after the epidemic had subsided in the locality."[55] Just as the WHO's World Influenza Program tracked the 1957 flu pandemic as it spread across national borders and through diverse populations, the WHO Veterinary Public Health division also attempted to assess whether the same virus would spread through animals. Veterinary services in thirty-three countries participated in the WHO-led animal survey, taking blood specimens from animals, freezing them or storing them at cold temperatures, and in most cases sending them to WHO influenza centers. Only a few positive serological results were obtained, but even this small number confirmed the hypothesis that the new strain could indeed infect pigs and horses as well as humans.

In a 1959 report summarizing the results, Kaplan and Payne argue that the survey findings, coupled with other evidence of the periodicity of pandemics and the "common country of origin (China)" of the 1889 and 1957 pandemics, suggest that "a parent influenza strain . . . resides in an animal reservoir on the mainland of China."[56] As I noted above, Zhu Jiming and other Chinese researchers had already come to the same conclusion. Once again, however, mainland China was one of the few parts of the world not covered by Payne and Kaplan's animal serological survey. Although C. H. Andrewes wrote to F. F. Tang of the Peking-based National Vaccine and Serum Institute proposing the collection of "sera from animals" in the "area in which the epidemic started"— even suggesting "the possibility that the Asian 'flu may have arisen from a reservoir in an animal in north China"—no sera were ever sent or reported to the WHO. During this period, the USSR, Poland, Czechoslovakia, and other Socialist Bloc countries all participated in the survey, but the People's Republic of China remained outside of the WHO's surveillance network.

Then, in 1968, history seemed to repeat itself. In July another new influenza variant was isolated in Hong Kong and soon spread across the entire world, causing what has been referred to as the "Hong Kong" flu pandemic. The *Times* of London first reported an outbreak of acute respiratory disease in southeastern China on July 12. Five days later, the Hong Kong Influenza Centre reported to the WHO an outbreak of influenza-like illness. After isolating viruses that appeared to be influenza but were antigenically distinct from the circulating A2 virus, Hong Kong sent the strains "as infected tissue-culture fluids on wet ice to the World Influenza Centre," and the WIC quickly confirmed a new variant strain.[57] The outbreak "adds a little more information to the oftenexpressed hypothesis that strains of influenza virus which have the capacity to spread widely and rapidly often arise" in southern China, wrote researchers from the WHO. However, China also remained a blind spot within the WIP surveillance system. Indeed, things were even worse than before, as China was in the midst of the upheavals of the Cultural Revolution. "Unfortunately," the WHO scientists continued, "contact between health authorities in China and other countries is even more difficult than in 1957 and it is impossible to obtain information on the possible origin or behavior of the epidemic prior to its appearance in Hong Kong."[58]

In addition to reaffirming the hypothesis that pandemic flu viruses arise in southern China, the 1968 Hong Kong flu pandemic also strengthened interest in the possible existence of an animal reservoir.[59] Since 1957, the WHO had continued to work with collaborating laboratories to develop what was becoming known as the "natural history of influenza" and later the "ecology of influenza." By 1969, much had been learned, but the major question still remained: "Are lower animals of any importance as a prime source of the major antigenic shifts in the influenza A group that cause the recurring epidemics and pandemics?"[60] Was there a "ghostly reservoir as yet undiscovered"?[61] To find out, the WHO's Martin Kaplan proposed a research program that would include the isolation of viral strains from wild animal species, continued surveillance of animal influenza in different countries using standard diagnostic kits, and laboratory research on the so-called recombination of influenza viruses. Of these, the last—though seemingly far removed from ecological questions—played the most important role in constructing an ecology of influenza. Recombination experiments rewrote the ecological relationships among human, animal, and bird species from inside the laboratory.[62]

VEHICLES AND VESSELS

During the 1930s viruses came to be seen as "prototypical models for hereditary particles" or something like "naked genes."[63] Used as model organisms, virus research prompted new forms of experimentation and paved the way for the movement of genetics research toward a molecular scale. Because these model organisms were also pathogens, intimate and frequent exchanges took place between "basic" research on genetics and "applied" research for public health, with the same viruses under the microscope and in the centrifuge. Influenza was one of the most important.

Frank Macfarlane Burnet, a researcher with the Hall Institute in Australia, had been a fellow of the National Institute of Medical Research in London during the years when Andrewes and colleagues discovered the influenza virus. When he returned to Australia, Burnet continued work on influenza, attempting to develop a vaccine to protect troops during World War II. His vaccine failed, but his immunological approach led him to study what caused changes to influenza virulence. To do so, he developed some of the first experiments in the genetics of the influenza virus, a research program that soon produced significant findings about viral genes.

Burnet's trajectory was complex and multidirectional. Although moving from the immediacy of public health to "basic" genetics research, his findings in experimental genetics almost immediately returned to transform the scope and scale of influenza as a public health problem. Burnet found that the influenza virus not only mutates, as was common to many viruses, but can also recombine: that is, two strains of virus placed in the same culture medium can exchange genetic material among themselves.[64] In 1959 he went even further to suggest that the high frequency of recombination indicates that influenza viruses must also undergo reassortment, the exchange of complete gene segments.[65] After conducting similar studies at New York City's Rockefeller Institute, Edwin Kilbourne argued that such reassortment could be a source of the genetic novelty that is required for the emergence of pandemic strains. If so, he speculated, animal reservoirs carrying different strains of flu may indeed play a key role in the emergence of human pandemics.[66]

But it was experiments by Robert Webster that intensified the implications of these findings when they eventually showed that "hybrid viruses" could be artificially produced in vivo as well as in vitro. Webster was born in New Zealand and trained in Australia during the late 1950s before establishing his permanent laboratory at St. Jude's Children's Hospital in Memphis, Tennessee. In the wake of the 1968 Hong Kong pandemic, Webster noted that the H and N antigens of the new variant A2 virus had cross-reactivity with viruses isolated from ducks and horses. He called for a new research program: "An explanation of the occurrence of human influenza viruses with animal or avian antigens should now be sought in a positive fashion. The tools are at hand for producing antigenic hybrids of human or animal influenza viruses in the laboratory and the next question we should try to answer is whether virulent antigenic hybrids can be produced in natural hosts."[67] Taking influenza viruses extracted from multiple species, Webster co-infected a single animal host with both viruses under laboratory conditions. Infecting turkeys with two distinct influenza viruses, for instance, Webster later isolated a hybrid or "new" virus that could infect and kill a chicken immunized against both "parental viruses." In another experiment, he showed that a human and a swine influenza virus could reassort inside of a single porcine host, which he later called a "swine mixing vessel."[68] After flu viruses infected an animal host, Webster showed they exchanged parts inside the animal organism, emerging transformed. As he explained in a 1970 letter, the experiments showed "the role of lower mammals and birds as vehicles in which genetic interaction between the influenza A viruses could occur."[69] In Webster's ecology of influenza, animals were becoming vessels and vehicles of virus reproduction and evolution, where before they were merely vectors of virus transmission. In a conclusive 1974 article, Webster and his coauthor, Charles Campbell, argued on the basis of these experiments that pandemic flu viruses most likely derive from the reassortment of human and animal influenza viruses, rather than the gradual mutation of seasonal flu viruses.[70]

Laboratory research was accumulating experimental results that appeared to expose the interspecies contacts and exchanges at the heart of influenza emergence. Still, experimental studies provided only "circumstantial evidence" of the natural history of flu pandemics, Webster and Campbell acknowledged. In order to verify that their experimental hypothesis held true "in nature," they suggested two possible routes of further research. The first approach would "detect recombinations of influenza viruses in nature prior to the emergence of the next pandemic strain," but they feared the "rarity of such an event" made this approach unlikely to produce convincing evidence. More promising, they argued, would be an "ecological approach" based on the establishment of a "bank of influenza viruses" isolated from wild birds, domestic poultry, and livestock. This was something like what Graeme Laver had done with wild seabirds in Australia and Dmitry Lvov was doing with a range of animal species in the Soviet Union.[71] Following the emergence of a pandemic virus, Webster and Campbell suggested, the new variant strain "could be compared with the viruses in the bank to determine if recombination plays an important role in the emergence of human strains in nature."[72] The ecology of influenza, as it was beginning to be called, would be constituted out of a bank of influenza viruses collected and stored in the laboratory.

INTO THE EPICENTER

In 1982 Geoffrey Schild, a virologist at London's World Influenza Centre, summarized a London conference meeting into a proposal for a WHO-led "Global Programme on the Ecology of Influenza." The proposal suggested that the ecology program could be modeled on the World Influenza Program's collection of human influenza isolates and would emphasize the importance of global scale. Schild recalled how the WHO developed the WIP because the "capricious epidemiology of influenza in man could be adequately studied only on a global basis" and suggested that the same global scale would be necessary for animal surveillance. The Programme on the Ecology of Influenza would also emphasize the collection of viral isolates, but from "non-human sources" of influenza. New projects included a survey sent to influenza researchers around the world, similar to Kaplan's 1957 survey, asking for the collection and submission of serological studies or isolated viruses on equine influenza.[73]

As we have seen, China had become a particular concern for global influenza surveillance for two reasons. First, historical evidence pointed to China as the origin of the viruses that caused the 1957 "Asian" and 1968 "Hong Kong" pandemics. As Webster's laboratory experiments accumulated evidence that an "ecology of influenza" enabled interspecies viral reassortments and the production of "new" influenza viruses, the spatial identification of China as an origin of pandemic viruses gained an ecological depth and scale. What was it about China's farmed ecology that encouraged the reassortment of influenza viruses? Was an animal reservoir somewhere inside China? Second, China's territory had been largely absent from the purview of even the most global programs of influenza surveillance because of the exclusion of the People's Republic from the United Nations and the politics of the Cultural Revolution. Questions about the origins of pandemic influenza were therefore lacking crucial data that could be obtained only inside mainland China.

By the 1980s, however, both science and geopolitics were in motion, traveling toward Chinese landscapes. In July 1971, socialist Albania and sixteen other nations proposed Resolution 2758 to the United Nations General Assembly, calling on it to "expel forthwith the representatives of Chiang Kai-shek from the place which they unlawfully occupied at the United Nations and in all the organizations related to it." This was not the first such proposal: Albania, stubbornly persistent, had proposed similar resolutions throughout the 1960s, and each time, equally stubborn, the United States had ensured that the proposals failed to muster sufficient support. But by the early 1970s, the lines across the political globe were taking new forms. Henry Kissinger had just been to Beijing, preparing the way for President Richard Nixon's landmark 1972 visit to China. Although the US still tried to find a diplomatic way for Taiwan to remain in the UN, its moderate proposal failed, and Albania's passed (Taiwan withdrew immediately before the passage of Resolution 2758). Over the course of the next few months most of the UN agencies, including the WHO and the FAO, had substituted the Beijing-based People's Republic for the Taipei-based Republic as the legitimate representative of China.[74]

Almost immediately following the inclusion of the People's Republic, the WHO began to organize study trips to send experts and officials to observe health care in China and initiate scientific exchanges. Many of these trips focused on topics like primary health care and China's celebrated "barefoot doctors." Pandemic influenza was a focus of a smaller number of WHO study trips. In September 1972, Webster and Laver traveled to China with a group of Australian medical doctors on an unofficial trip because they "hoped that the Chinese might have information about animal and avian influenza viruses occurring in China." The trip was disappointing, despite the "enthusiastic friendliness" that the Australians encountered from officials, hospital staff, and the general population. The itinerary was "pretty inflexible"; as Laver reported to the WHO, "Most of our time was spent visiting hospitals where we saw traditional Chinese medicine as well as Western medicine being practiced," and the highlight seemed to be observation of acupuncture anesthesia during surgery, hardly of scientific interest to their pandemic research. As Laver recalls,

> We did not find out much about influenza. In one city we saw a group of pigs, wallowing in the mud. We asked the Chinese if we could take samples of the pigs' blood to see if any antibodies could be found. There was a good deal of resistance to this request but after much haggling we were allowed to bleed one pig. Since a single sample does not do a great deal for the statistics we asked if we could have some more. Came the answer: "In China today all pigs are equal; you have your sample, be satisfied."[75]

Webster and Laver were able to contact virologist Zhu Jiming, however, and they spent "a whole day" at the National Vaccine and Serum Institute (NVSI), where Zhu ran a laboratory on virus research at the time. Zhu helped translate their talks to the NVSI, provided information about China's vaccination and surveillance programs, and even shared isolates related to the 1957 pandemic strain.[76] They also observed "how primitive Chinese laboratories had become during the dark days of the Cultural Revolution; no deep freezers, no equipment for storing live viruses for long periods of time." Throughout their visit, they were "unable to find out anything about influenza viruses infecting animals or birds."[77]

In 1981 Webster applied for funding from the U.S. National Institutes of Health for a conference to be held at the Institute of Virology in Beijing on the "origin of influenza viruses."[78] In his proposal for the workshop, Webster notes that several pandemics are widely believed to have originated in China—including the 1957, 1968, and 1977 pandemics—and he suggests that this information deserves to be "examined in the light of recent knowledge of influenza and its ecology in animals." In other words, the planetary hypothesis of China as origin point needed to be redrawn based on the ecological inferences coming out of Webster's laboratory. Yet to do so would also require forging new relations of scientific collaboration, Webster acknowledged. "The question of 'why China,'" he explained, "should be reviewed in conjunction with Chinese scientists." After all, "Chinese scientists are in a unique position to resolve the intriguing question of why so many human epidemic strains have originated in their country." As a result, the aims of the workshop included not only immediately scientific ones but also the training and development of a Chinese infrastructure for viral surveillance. The workshop would, he proposed, "stimulate the use of modern molecular biological methods in China so that an explanation to the intriguing question of why many 'new' pandemic strains arise in China may be forthcoming; such information could provide a basis for effective control in the future." Equally important, the workshop would provide a venue to "establish personal and scientific contact with a number of virologists in China who have previously not been contacted" but had expressed interest in "collaborative studies with overseas virologists."[79]

Laver co-organized the workshop, while Zhu Jiming, newly appointed as director of China's Institute of Virology, conducted local coordination in Beijing. Papers presented ranged from molecular studies of the structure of antigenic proteins to immune response, and included four papers on "viruses from birds and lower mammals." A team from the Institute of Virology—led by Guo Yuanji and including Zhu Jiming—presented a paper on what they called "influenza ecology in China." The paper reported the results of "virus surveys" that the institute conducted on domestic ducks, feral ducks, and pigs. These surveys isolated a broad range of influenza viruses, including a "great diversity of subtypes of influenza A," from domestic ducks. Guo and colleagues concluded that these findings were "consistent with the hypothesis" that avian and pig viruses "play an important role in the origin of new subtypes" of human flu.[80]

Another conference on the ecology of influenza was held in Hong Kong, then a British colonial city, the same year. The meeting was chaired by C. H. Stuart-Harris, who had been a research assistant at the British Medical Research Council (MRC) when influenza virus was first isolated, and had made several important contributions to flu research himself. The rest of the dozen or so participants came from veterinary, public health, and virology institutes in Hong Kong or China (PRC), including Guo Yuanji from the Institute of Virology and G. Z. Shen from the Health and Anti-Epidemic Station in Guangzhou. At the meeting, the group examined "some features of human and animal life in Hong Kong and the southern regions of China where epidemics of influenza A caused by novel surface antigens—hemagglutinin (H) and neuraminidase (N)—have arisen and have spread as pandemics throughout the world."[81]

Although Stuart-Harris chaired the meeting, the driving force behind the discussion of the ecology of influenza was Kennedy Shortridge. An Australian by nationality, Shortridge moved to the microbiology laboratory in Hong Kong—at the very edge of mainland China's territory—soon after the so-called Hong Kong flu of 1968, inspired by the hope of detecting the onset of the next pandemic virus.[82] Once there, he began to collect influenza viruses, primarily from poultry, and type them in his lab. The laboratory collected viruses during a weekly or fortnightly influenza surveillance program at a Hong Kong "poultry dressing plant" that sourced chicken, geese, and ducks from farms both in Hong Kong and in the southeast of the People's Republic. Additional viruses were collected from duck feces and pond water at a domestic duck farm in Hong Kong.[83] The surveillance program isolated a wide variety of influenza A viruses, successfully typing 46 out of the 106 HA/NA protein combinations possible under the existing nomenclature system. Hong Kong proved to be an "extremely fruitful source of influenza viruses," Shortridge wrote.[84]

In total, at least three conferences on the "ecology of influenza" took place that same year. The ecological focus of attention, and the abrupt geographical movement of research programs toward China, prompted a somewhat disgruntled response from Alan Kendal of the U.S. CDC. "At first thought," Kendal scrawled on CDC Influenza Center notepaper after reading Webster's conference proposal, "it is hard to see why WHO should support the same old group of people to come together in China and use fallacious arguments (or non-arguments) to involve China as the source of a new strain."[85] However, Kendal's skepticism overlooks the fact that the discussions at these conferences were not exactly repeating "the same old" arguments about China as an origin of pandemics. In fact, Shortridge, Webster, and others were beginning to build connections between two distinct ideas: on the one hand, China as the geographical source of influenza pandemics, and on the other, laboratory understanding of the interspecies communications that cause the reassortment of new viruses. Together, the two ideas pointed to China as the *ecological* birthplace of pandemic influenza viruses.

In a report on the Hong Kong discussions submitted to the *Lancet* that explicitly linked these two ideas, Shortridge and Stuart-Harris proposed that southern China might be an "epicentre" of influenza pandemics. Describing the agrarian landscapes of southern China, they argued that the ecology of the region promoted "exchange of viruses between animals and man": "The closeness between man and animals could provide an ecosystem for the interaction of their viruses." Of particular concern was the human-cultivated, "age-old" ecosystem of rice paddies, waterways, and poultry and pig farms: "The densely populated, intensively farmed area of southern China adjacent to Hong Kong is an ideal place for events such as interchange of viruses between host species."[86]

The hypothesis of the influenza epicenter transposed experimental facts produced in the laboratory onto the ecologies and environments of southern China. As I discussed above, Robert Webster's laboratory research had demonstrated the important role of interspecies contacts in the reassortment of viruses. The co-infection of a host by viruses originating from two species allowed these viruses to exchange genetic material, producing new variants of flu. Shortridge claimed, quite simply, that the relations among species artificially constructed in Webster's laboratory were the stuff of everyday life along the Pearl River Delta. "In the villages," he and Stuart-Harris wrote, "it is common to see ducks, geese, and chickens running loose in proximity to pigs and water buffaloes and to see small children playing in the environment." If interspecies contacts led to viral reassortments, and viral reassortments caused pandemics, then for these reasons China's densely populated and intensively cultivated southern region was probably a "point of origin" for flu pandemics.[87] No longer simply showing that China was where influenza pandemics started from, Shortridge and others were beginning to answer the questions posed by Zhu Jiming: How and why did they begin there?

Through research on the influenza virus in the laboratory, knowledge of the virus began to point beyond the virus—toward specific regions like southern China and toward ecologies of poultry farms, pigsties, and waterways. In the next chapter we will see how the effort to answer these questions brought the science of pandemic flu out of the laboratory, displacing the concept of an ecology of influenza in the process.

In crafting the influenza epicenter hypothesis, Kennedy Shortridge deployed environmental and ecological tropes of disease and pathology, but his experimental system remained bound within a laboratory scale. In many respects, Shortridge's practice resembles the classical bacteriology of Louis Pasteur, as presented in Bruno Latour's account of the laboratory as a "center of calculation."[1] Shortridge, like Pasteur, ventured out to farms and slaughterhouses to collect biological material, then brought the material back to his laboratory to be translated by inscription devices, including hemagglutinin inhibition tests.[2] Using the language of phylogenetics, he built a classified and ordered archive of samples typed according to their hemagglutinin and neuraminidase proteins: HINI, H7N2, H9NI. Drawing from this *bank* or *profile* of influenza ecology, Shortridge built narratives about the ecological conditions of pandemic emergence.

Theresa MacPhail points out that alongside rapid advances in molecular biology, claims about the origins of pandemic flu increasingly drew from studies of the genetic relations between viruses: "Phylogenetic *relationships between strains* are analyzed to produce knowledge about the 'ecology' of influenza viruses."[3] For instance, during the 1990s Robert Webster and others identified wild aquatic birds as the reservoir of influenza viruses by comparing genetic sequences from a human flu virus with flu viruses sampled from aquatic birds.[4] Yet when I began my fieldwork in early 2010, I found that scientists working inside China had an understanding of the "ecology" of influenza very different from Webster and Shortridge's viral banks and phylogenetic charts. The popularity at the time of the phrase "One Health," often used to refer to a broader approach to flu research demanding collaboration across disciplines, was one sign of this nonvirological understanding of the influenza problem.[5] Talks on One Health frequently took place, the United Nations' China-based offices began a new "sub-working group" on the "human–animal disease interface" in 2010, and this working group organized a two-day "One Health China Event" in Beijing the following year. As a researcher explained to me, many of the people working on flu in China were "trying to embed molecular processes in the larger ecology." This chapter aims to specify the contours of this emerging eco-virology of influenza and its differences from an earlier virus-centered ecology. In what way is the molecular scale of viruses and genes embedded in a larger-scale ecology?

Of course, research on the ecological conditions of disease long predates the term *One Health*. Disease ecology has been described as a "minor tradition" in twentieth-century infectious-disease research.[6] And as we have seen, flu researchers from Robert Webster to Zhu Jiming to Kennedy Shortridge had developed what they called an "ecology of influenza" since the early 1970s. This chapter specifies the epistemological displacements that have taken place in recent flu research at the epicenter, despite remarkable continuities in research programs over the past fifty years: continuities that include the use of terms such as *ecology*. To uncover this displacement, I focus attention on a single truth claim that is essential to the identification of southern China as the epicenter of pandemic flu viruses: the hypothetical role of free-grazing ducks in the emergence of new viral strains. Tracing how Kennedy Shortridge articulated this claim with experimental evidence in the 1980s, I then follow the reassertion of the claim in contemporary pandemic flu research. By following this reiteration of the same claim, the epistemological displacements of the *forms* for making truth claims—the experimental systems, to use Hans-Jörg Rheinberger's term—are made visible. At stake, as we will see, are two distinct concepts of ecology: the first built upon the phylogeny of viruses and the second constructed through spatial models of landscape and species interaction.

In criticizing how discourses about pandemic epicenters construct a geography of blame, anthropologists have highlighted how these scientific studies rely on unscientific tropes of exotic environments and "backward" traditional cultures.[7] In this chapter I approach Shortridge's ecological claims in a different manner. Tracing dramatic changes in the form and scale of scientific work on the epicenter, I show how Shortridge's anecdotal claims about environment have been turned into research objects in spatial ecology. This does not mean that science freely escapes history or ideology. On the contrary, it means that the trajectory of scientific change is shaped by the displacement of ideological pasts.

TWO EPISTEMOLOGIES OF ECOLOGY

At the center of Shortridge's viral ecology of influenza is a claim about ducks, not viruses. By 1982, Shortridge had spent over five years collecting, analyzing, and archiving influenza viruses from poultry and other livestock. He found that Hong Kong was full of flu viruses, but he also found that these viruses weren't everywhere. In one article Shortridge reported with evident pride that he was able to isolate and type nearly half of the 106 hemagglutinin and neuraminidase protein combinations possible under the existing classificatory system. He also emphasized an even more unexpected finding: he isolated all but one of these combinations from a single species: domestic ducks. Overall, Shortridge consistently found sixteen times as many influenza viruses in ducks than chickens in each of the five years that he looked for viruses. Shortridge concluded that the results demonstrated "the importance of ducks in influenza ecology."[8]

Before he reported his results, there was no consensus among influenza researchers about the role of domestic waterfowl in the emergence of influenza pandemics. Although some virologists speculated that ducks could be a reservoir of viruses, and a source for the viruses that cause human pandemics, others suggested that ducks may receive viruses *from* humans *after* pandemics. After "unearth[ing] a seemingly bottomless reservoir of viruses" in southern China's ducks, including many viruses that had never been isolated from humans, Shortridge argued that "domestic poultry might be donors of viruses adaptable to human hosts rather than recipients."[9] In a paper presented to the "Origin of Pandemic Influenza" conference held in Beijing in 1982, Shortridge included a diagram that, he claimed, described the "generalised sequence of transmission for the emergence and spread of a pandemic virus." To craft this hypothetical sequence, he drew from the growing bank of influenza viruses he had assembled, a bank that Frédéric Keck has compared to a museum archive.[10] In the diagram, the virus leaps from wild waterbirds to domestic ducks to humans and back again, before finally ending in the domestic pig (unconventionally considered in this diagram model as a dead-end host).[11] "Since duck raising may be fundamental to the emergence of pandemic viruses," Shortridge concluded somewhat pessimistically, "then it would seem that the control of influenza at source may be an impossibility given the continuity of the farming practices in the region and the high density of the rural population."[12]

In developing these findings into the hypothesis of the pandemic epicenter, however, Shortridge did not restrict himself to statements about the frequency of detected viruses. A close reading of Shortridge's classic article, "An Influenza Epicentre?" (coauthored with C. H. Stuart-Harris), reveals the assertion of two distinct kinds of truth claim. One set of claims presents the facts constructed in Shortridge's experimental system: concise virological results, often supported by reference to more technical articles published by Shortridge or others. These strictly virological statements are juxtaposed, however, with broader claims about the ecology and ethnology of southern China. For instance, Shortridge and Stuart-Harris highlight what they call an "age-old" agricultural practice: the use of ducks for manuring and protection of the wet-rice paddy. They adopt an almost ethnographic style: "[Farmers] make use of the food preferences of domestic ducks which help protect the growing rice from insect and shellfish pests and carry out weeding, intercultivation, and manuring. This practice reduces farmers' dependence on chemical insecticides, herbicides, fertilisers, and mechanical farming aids and provides a close bird/water/rice/man association that varies with the seasons of rice-growing."[13]

The hypothesis of the pandemic epicenter relies as much on these ethnological claims as on the results of laboratory experiments. Put another way, if we ask how virological findings are transposed onto southern China's landscapes, it becomes clear that in this case ethnological statements—and not virus sampling—are the crucial instrument of transposition. Yet, these ethnological claims never became the scientific objects of Shortridge's experimental system. Instead, Shortridge supports his ethnology with a diverse array of rather more anecdotal evidence: personal communications, his own travel experience, unpublished papers, and even Joseph Needham's magisterial (but historical) *Science and Civilization in China*.

Inside the hypothesis of the influenza epicenter, therefore, are two kinds of facts with different epistemological standing. Shortridge's laboratory research produces a lot of facts about the role of ducks in what he calls "the ecology of influenza," but his viral profiles, on their own, do not support the claims he makes about the agricultural ecology of ducks and duck farming in southern China. Viruses are studied with analytic precision, but the ecologies of farm animals and rice paddies are described in more "metaphoric" terms, to adopt Warwick Anderson's distinction.[14] Historian of biology Georges Canguilhem describes this second type of truth claim as a *scientific ideology* and suggests that such ideologies play a crucial role in the historical trajectory of scientific change. As historians have shown, scientists rigorously define their objects, producing what Rheinberger has called "epistemic things" that are fundamentally distinct from everyday things. Yet these epistemic things do not, and cannot, exist in a pure culture, in absolute isolation. Rather, they are supported by a matrix of linkages to politics, modes of representation, and economies of value. These linkages ultimately provide much of the motor driving scientific change: pointing toward new research problems, suggesting the significance of research findings, applying research questions to new domains. Although sometimes derided as hype, these para-scientific linkages are, even more than experimental systems proper, the manner in which scientific "futures [are] carved out of the present."[15]

The concept of scientific ideology illuminates a model of scientific change that is neither as radical and absolute as a paradigm shift nor as linear and gradual as "development-by-accumulation."[16] A scientific ideology, according to Canguilhem, both precedes and follows upon the constitution of a scientific object or epistemic thing. A scientific ideology is not a fraudulent science, nor is it antiscience: indeed, it has the "explicit ambition to be a science."[17] It is more like a virtual cloud suggesting the outlines of a near future that has yet to become actualized in a scientific experiment.

More recently, Rheinberger and Staffan Müller-Wille describe the working of scientific ideology in their cultural history of heredity. Nineteenth-century ideas about heredity provided a background of interests and motivations that guided the experimental research that ultimately constructed the "gene" as a scientific object. Concern about heredity first emerged in areas such as animal breeding or legal inheritance. Early scientific investigations largely derived from interests and anxieties about race in the late nineteenth century. Initially, these ideas about heredity were only loosely linked to experimental findings, at best. Yet, as Rheinberger and Müller-Wille stress, the objectification of the gene would not have taken place without them.

Even when the ideologies about the imagined significance or potential of a scientific project don't come true, something new is created in the process. What is important from a historical point of view is that this pathway toward the new was in part *pushed* by those ideologies in a certain direction. The scientific objectification of ideology is not a process of fulfillment, in which the imagined future is actualized. Rather, Canguilhem explains, objectification is a process of *displacement*: "Scientific ideology does indeed stand over [*superstare*] a site that will eventually be occupied by science. But science is not merely overlain: it is pushed aside [*deportare*] by ideology. Therefore, when science eventually supplants ideology, it is not in the site expected."[18] Canguilhem is playing on the idea of superstitions (*superstare*) as ideas that "stand over" future scientific sites, but also challenging the idea that science follows a linear path of progress out of superstition, as if simply replacing or overlaying true knowledge on top of false belief. Instead, scientific ideologies lay out a direction of inquiry, but the actual results of that inquiry are pushed in unexpected directions.

In 1997 the emergence of the novel H5N1 highly pathogenic avian influenza virus in Hong Kong, and the attribution of several human deaths to H5N1 infection, was at first widely seen as a confirmation of Shortridge's hypothesis. Shortridge himself reiterated that "China is the principal reservoir of influenza, and southern China is the influenza epicenter."[19] Faced with what appeared to be a pandemic warning, Hong Kong's veterinary administration slaughtered all poultry in the territory. The virus seemed gone. The world was saved.

But then the virus kept reemerging: isolated from geese imported for slaughter in Hong Kong (1999), in ducks from China's Guangxi Province (1999), in geese sold at Vietnam's poultry markets (2001), and again in a duck in Guangxi (2001).[20] As Les Sims, Hong Kong's assistant director of agriculture quarantine, told me, "Virus continued to circulate in China all through from 1996 to 2004 and the absence of reports of disease does not reflect the true infection status. . . . It is clear from basic biology that disease must have been occurring in the mainland but for whatever reason was not being reported." In 2003 the H5N1 virus reemerged in Hong Kong before spreading rapidly and widely throughout Southeast Asia. The repeated emergence of H5N1 and related viruses set the stage for the displacement of Shortridge's experimental system. His anecdotal or ideological claims about the agricultural ecology of southern China were pushed aside by a new set of experimental systems that made the ecology of the epicenter into a scientific object of its own.

FROM VIRUS TO LANDSCAPE

The virus that reemerged in 2003 was genetically somewhat distinct from the 1997 virus, showing that the virus had accumulated a series of mutations in the intervening years.[21] More important than this molecular change, however, were changes in the geographic extension of viral infection. By 2004, the virus had spread far beyond Hong Kong, with large outbreaks appearing almost simultaneously in the poultry farms of Thailand and Vietnam and ravaging the flocks of smallholders and industrial giants alike. I first realized the importance of this newly extended viral geography in conversation with Robert Wallace, a self-described "phylogeographer" whom I met in Beijing. In his work, Wallace tries to link the molecular phylogenetics of virus strains with the spatial geography of their isolation, developing a narrative of viral origins and emergence.[22] He told me that even without mutations to the molecular structure internal to the virus, a changed epidemic geography can *itself* transform the capacity for transmission, offering new challenges and opportunities for control. "Once the level of coordination changes, spatially," he explained, "the nature of the problem changes." For instance, when Ebola spread out of remote villages and into large urban areas in West Africa in 2015, the mode of infection changed even though there was no change to the virus. The ecology of host networks, not the virus, had mutated.[23]

As avian influenza A H5N1 spread out of the pandemic epicenter and beyond Hong Kong, the mode of infection did not change much, but the epidemiological problem certainly did. As it spread through Southeast Asia and beyond, and as it persisted in poultry populations, the virus began to take on a geography. Kennedy Shortridge's laboratory had perched at the frontier of the epicenter to detect the onset of a pandemic before it emerged. The objective had been to collect and classify a bank of influenza viruses so that any new strain could be identified and compared with the viral archive. Now, as the virus spread, geography and landscape became new research objects for understanding the ecologies of an emerging disease.

In the rest of this chapter I chart this process of displacement by following the reiteration of one of Shortridge's most significant claims about the influenza epicenter: the "importance of ducks in influenza ecology."[24] I do so by discussing the work of spatial ecologist Marius Gilbert and his collaborators, a team of researchers linked to the Food and Agriculture Organization Emergency Center that I followed during my fieldwork in China. By restating Shortridge's claim in terms of a very different experimental system, Gilbert did not simply add new facts to the store of knowledge about pandemic flu; he also displaced the phylogenetic concept of the influenza epicenter. He pushed aside anecdotal claims about the landscapes of the influenza epicenter by turning them into objects of a spatial ecology.

Gilbert completed a PhD in agricultural and allied biological sciences from the Free University of Brussels in 2001, and he is now affiliated with the same university's Center for Biological Control and Spatial Ecology. Spatial ecology derives from a simple but fruitful principle: "In nature, living beings are distributed neither uniformly nor randomly. Rather, they are aggregated in patches, or they form gradients or other spatial structures."[25] Spatial ecology attempts to identify the ecological rules that explain the distinctive pattern of population distribution for a particular organism. In Gilbert's doctoral dissertation, he modeled the invasions of the bark beetle *Dendroctonus micans*, and in subsequent work analyzed the spread of the horse-chestnut leaf miner (*Cameraria ohridella*), a leaf-mining moth, in the United Kingdom.

Gilbert's recent work on avian influenza transposes the tools and equipment of spatial ecology from pest invasions to viral epizootics. As a brief biographical statement on Gilbert explains, his "work initially focused on the spatial ecology of invasive insects. In the last 10 years, he became interested in the way concepts and methods usually applied to invasion ecology could be used to improve our understanding and modeling of epidemiological study systems, and started working on several animals [*sic*] diseases such as Bovine Tuberculosis (BTB), Foot and Mouth Disease (FMD), and Highly Pathogenic Avian Influenza (HPAI)."[26] In a coauthored methodological statement published in 2004, Gilbert wrote that key questions about emerging diseases ("How did a pathogen enter a human population? What are the characteristics of potentially invasive pathogens?") could be effectively answered with the tools of spatial ecology: "These questions relate to the adaptation of a pathogen to a new host, and can be addressed by comparisons between emerging diseases and invasive plant and animal species."[27]

Gilbert told me that his first opportunity to study HPAI "on the ground" was in Thailand. After reemerging in Hong Kong, HPAI H5N1 struck Thailand in January 2004, roughly concurrently with similar outbreaks in Vietnam. A second outbreak, known in Thailand as the "second wave," began in July 2004 and grew in severity and scale throughout the summer. At first, the epizootic appeared to catch the Thai government by surprise. The Ministry of Public Health (MPH) and the Ministry of Agriculture and Cooperatives (MAC) fought over jurisdiction and policy, hindering efforts to control the spread of the virus. By the end of September, however, they began to cooperate with the Department of Livestock Development (DLD) on a surveillance and control program that they called the X-Ray.[28]

With the X-Ray, Thailand established an unprecedented *active* surveillance program for avian influenza. Most influenza surveillance systems are passive—that is, they collect and collate ad hoc reports from laboratories, physicians, or veterinarians. The WHO's World Influenza Program exemplifies such passive surveillance. The central reference laboratories do not actively search for flu viruses across the globe but in fact depend on reports of outbreaks and submissions of viral specimens from other laboratories around the world. Active surveillance, as its name implies, involves building infrastructure to actively search out disease across a particular territory or population. The Thailand X-Ray aimed to provide just such a "composite picture of HPAI situation in Thailand" during a one-month period. The program established thousands of surveillance and rapid response teams (SRRTs) that built upon and transformed existing public health infrastructure, most notably adapting more than 700,000 existing village health volunteers (VHVs) into a grassroots surveillance system.[29] The VHVs investigated any unusual poultry or human infection within their villages, sending reports of outbreaks to the SRRTs for subsequent field sampling and laboratory verification of the causal pathogen.

In addition to facilitating the Thai government's control and eventual eradication of the virus, the X-Ray also produced a rich data archive. During the first month of the X-Ray, samples were taken from 150,648 birds, of which 724 tested positive for H5N1. The X-Ray also included a "door-to-door" poultry survey, in which VHVs collected detailed information from farmers about the breeds and size of their flocks.

The scale and quality of this data archive presented a unique opportunity for applying spatial ecology to the problem of an emerging disease. In particular, Gilbert found the data useful because the data were derived from active, rather than passive, surveillance. Although passive surveillance systems are suitable for the sentinel detection of emergent strains or unexpected outbreaks, their use for spatial modeling is very limited. Passive surveillance provides no information about geographic locations that did not report strains or outbreaks, and therefore is spatially biased. Active surveillance, by contrast, seeks out disease or pathogen in every location included under the surveillance system, whether or not disease was reported there, in order to confirm the presence or absence of the pathogen. As a result, a more complete spatial picture is created.

The X-Ray, in this case, constructed a database of diagnostically verified outbreak reports, including the geographic location and the date of all poultry outbreaks during a one-month period. At the same time, because surveillance teams were deployed in all communes across Thailand, the absence of reports of disease had also been verified to be the absence of actual disease. The quality of this database allowed Gilbert to construct the first spatial model of the "invasion" of avian influenza H5N1.

In the model, Gilbert sought to determine the spatial associations between laboratory-confirmed presence of the H5N1 virus and a series of poultry and environmental variables, designated as risk factors. The outputs of the models, as presented in the resulting 2006 publication, include a series of six maps of Thailand. Two of the maps show the distribution of highly pathogenic avian influenza outbreaks in chickens and ducks. The subsequent four maps depict the population distributions for broilers and layer hens, native chickens, free-grazing ducks, and meat and layer ducks. The counterintuitive results, mathematically explored with stepwise multiple logistic regressions,[30] are actually visible to the naked eye with a careful examination of the maps: the spatial pattern of outbreaks in chickens closely matches the locations of free-grazing duck populations, rather than the spatial distribution of chicken populations. As Gilbert and his coauthors explain, "Although most HPAI outbreaks during the second epidemic in Thailand occurred in chickens, the spatial distribution of these outbreaks does not correspond to areas with high densities of chickens. . . . Instead, the distribution pattern suggests an important role of free-grazing ducks in rice paddies as in the central plains of Thailand."[31] In a joint report prepared for the FAO and Thailand's DLD, Gilbert and coauthors stated their argument more forcefully: "The implication is that ducks may play a role in generating chicken outbreaks, but not the other way round."[32]

By pointing to the importance of ducks in the ecology of highly pathogenic avian influenza, Gilbert's study reaffirms Shortridge's decades-old claim. Precisely because of this apparent continuity, however, Gilbert's work exposes the displacements at stake in the turn from phylogenetics to eco-virology. For by shifting Shortridge's virological results onto a spatial register, Gilbert opened up new perspectives and new approaches for inquiry.

Gilbert's research objects are situated at a landscape scale rather than molecular scale. This displacement is particularly visible when Gilbert reiterates one of the anecdotal claims about "traditional" agriculture that is at the center of Shortridge's influenza epicenter hypothesis—the practice of allowing ducks to forage in rice paddies: "Traditional free-grazing duck husbandry in Thailand is characterized by the practice of frequent rotation of duck flocks in rice paddy fields after the harvest, in which they are moved from 1 field to another every 2 days to feed on leftover rice grains, insects, and snails. Duck husbandry involves frequent field movements of flocks that are brought together in shelters often located within villages; with marketing of live birds and eggs extending beyond villages, apparently healthy ducks may play an important role in virus transmission, which explains the observed spatial pattern of HPAI."[33] The ethnographic tone of this passage resembles the anecdotal descriptions of a "close bird/water/rice/man association" proposed by Shortridge and Stuart-Harris in their article on the influenza epicenter. Yet Gilbert, unlike Shortridge, began to turn this ideological claim into a scientific object by developing a series of mechanisms for analyzing and quantifying the relationship between ducks and rice. Along with collaborators in Thailand, for example, he began to explore Thailand's rice-production statistics. Unexpectedly, the results did not show a simple correlation between rice farming and free-grazing duck populations. Rather, he found that free-grazing duck populations tended to be larger in areas that cultivated multiple crops of rice each year. In areas where farmers grew two or more crops on the same field, the "year round availability of post-harvest rice paddy fields" seemed to make free-grazing duck farming more likely.[34]

Nor did Gilbert stop with the claim that duck farming is linked with the agricultural practices and landscapes of the region. By showing that wet-rice-paddy cultivation (or more specifically, double-cropped rice) is linked to free-grazing duck populations, he argued that the one could analytically *substitute* for the other. And this substitution opened up a radical shift in the technical equipment of the experimental system, making possible the study of influenza from a new viewpoint: from seven thousand kilometers overhead.

"With the strong association between free grazing ducks and HPAI outbreaks," Gilbert wrote, "and also between rice and duck farming systems, an option that presented itself was to explore the application of remote sensing as a risk assessment decision support tool for HPAI."[35] Although initially used in earth sciences and climate or atmospheric research, remote sensing gained purchase in ecological research during the 1980s and 1990s. More recently, the use of satellites to monitor landscape drivers of disease emergence or vector movements—a field known as "tele-epidemiology"—has grown in prominence. In both cases the expansion in the use of satellite systems and remote sensing has paralleled the growth in commercially available imagery and better image resolution.[36]

In November 2005, Gilbert submitted a proposal to the FAO for funds to investigate whether satellite imagery could be used to predict the distribution of rice cultivation and whether these predictions could be used to assess avian influenza risk. As Chunglin Kwa has shown, the use of remote sensing often transposed the scale of ecological research away from "local ecologies" and field studies toward regional or "mesoscale" landscapes.[37] In this case the use of satellites as a new research instrument consolidated the shift from the molecular scale of viruses to a much broader landscape scale.

Gilbert sought the assistance of a research team led by Xiangming Xiao, a geospatial analyst who was already studying the use of satellite optical sensor data to predict rice-crop distribution. Xiao was born in China and has a degree in plant ecology from the University of Science and Technology in Beijing, but did his advanced graduate studies in the United States. He is now a professor at the University of Oklahoma and the director of Oklahoma's Center for Spatial Analysis. However, he continues to conduct much of his research on China's landscapes and in collaboration with Chinese research institutes. In a 2005 paper, Xiao and coauthors reported on the use of image data collected from the Moderate Resolution Imaging Spectroradiometer sensor on NASA's Terra satellite to map paddy-rice agriculture in southern China. Because the production of each crop of rice involves a flooding stage at the time of planting, the paper reports the identification of a spectral "fingerprint" (involving temporal shifts in the quantitative indices of vegetation and moisture detected by the satellite) that corresponds to the flooding of the paddies. By observing this fingerprint, Xiao's team claims to be able to map the geography of double-cropping rice.[38]

At first, Gilbert and Xiao used these remote sensing techniques to demonstrate the relationship between free-grazing duck populations (as recorded in the X-Ray survey) and rice-cropping patterns.[39] This was a preliminary step, a proof of concept, so that subsequent studies would now be able to use remote sensing of rice cropping as a substitute for free-grazing duck populations. Another study showed that the distribution of double-cropped rice, as identified by the satellite algorithm, was closely correlated with HPAI outbreaks in both Thailand and Vietnam. The authors point out that the substitution of rice cropping for poultry statistics offers important benefits to the experimental system: "An applied result of this study is that the distribution of rice cropping intensity can readily be established at any time and be used to complement traditional duck census data. Remote sensing data are available at a much greater spatial and temporal resolution than traditional censuses, thus allowing a fine-scale risk mapping."[40]

Poultry censuses are notoriously inaccurate: they are conducted by government agriculture departments, which are often inclined to inflate the numbers; they are conducted at most once per year; and they do not necessarily distinguish types of bird. Free-grazing ducks, in particular, are poorly captured in such official statistics: free-grazing is a difference of farming practice rather than species. Satellite mapping of rice cropping therefore produces a more "generalizable" risk factor than survey data on free-grazing duck distribution, as Gilbert explained in a 2011 conference presentation I attended in Beijing.[41]

Julien Cappelle, a PhD student with Gilbert in disease ecology, drew attention to the significance of satellites in marking out the specificity of contemporary spatial ecology and epidemiology. Much of the underlying methods of spatial analysis can be traced back to the nineteenth century, to early heroes such as John Snow. "What's new are the tools," especially remote sensing and satellite telemetry, Cappelle told me: "Satellite images, that's been around for a while now, maybe forty, fifty years, but at the beginning it was really military use only. Now you've got a large range of satellites that are really here to observe the earth, and that produces a lot of images. From these images you can actually get some environmental indicators." With Xiao's remote sensing system, he explained, you could map rice-cropping intensity at a landscape scale—or put another way, you could map the *variation* in the distribution of what Short-ridge had called bird/water/rice/man associations across a landscape.

Shortridge had sampled viruses, then broke them apart to understand their antigenic and genomic characteristics. Based on this work at a molecular or protein scale, he constructed hypotheses about the ecology of the pandemic epicenter. Gilbert, Xiao, and others now used satellites orbiting the Earth to document the spatial distribution of landscape and ecological features in order to better locate and differentiate the geography of viral emergence. Avian influenza was gaining a landscape topography, and soon this topography would begin to displace the hypothesis of the pandemic epicenter.

GAINING ENTRÉE

In 2009 Gilbert and Xiao co-organized the "International Workshop on Community-Based Data Synthesis, Analysis and Modeling of Highly Pathogenic Avian Influenza in Asia." Phylogeographer Robert Wallace recalls that the Bangkok workshop took place in a large room with four long tables arranged in a square formation. This spatial arrangement differed from the typical scientific conference, which is oriented toward the reporting of results for peer evaluation and where a podium and PowerPoint screen face rows of straight-back chairs. Many of the participants did not know one another, and they came from widely disparate fields: ecological modelers and wildlife veterinarians, socioeconomists and molecular biologists, epidemiologists and policy analysts. The square table created an architecture of familiarization for confronting the unfamiliar.

Wallace found a "meeting of minds" where, despite diverse backgrounds and different experimental systems, everyone was "looking for something else." The participants shared a certain antipathy toward laboratory and virological "reductions" of influenza epidemiology, Wallace remembers, and they all sought to "embed molecular processes in the larger ecology." As should be clear from my discussion of Gilbert's research, this search did not by any means involve a rejection of virology or the germ theory. Rather, the ethos of looking for "something else" reflects a hope that laboratory and molecular knowledge about viruses could be situated within a spatial and multispecies ecology of viral emergence. Viral phylogenetics had dominated research on influenza since the 1970s, extending its purview to make claims about even the ecology of influenza. The epistemic community taking shape in Bangkok would, over the next five years, propose a series of methods that resituated molecular phylogenetics within a broader "eco-virological approach."[42] The study of emerging infectious diseases has famously been described as an effort to monitor "viral traffic," that is, to observe and characterize the spillover of viruses from one species to another. Shortridge's laboratory had even functioned as a "sentinel" to keep watch over traveling viruses.[43] Now Gilbert and others aimed to build a very different device: a map of the highways and bridges of viral traffic.

The pandemic epicenter loomed in the shadows of the gathering. During the Bangkok meeting "China was a big question mark," Wallace reminded his colleagues at a subsequent meeting a year later. As Wallace explained to me, "What came out of that discussion in Bangkok . . . is that a lot of the new H5N1, various, substrains, and new avian influenza were coming out of southern China, and yet we don't have entrée into that . . . and so there was a sense of 'Oh, we're talking about everything but the most important thing of all on the table which is what's going on in China.'" The epistemological displacement of the "ecology of influenza," from laboratory phylogeny to spatial ecology, now drove a spatial displacement of global health projects, biosecurity interventions, and experimental systems *into the epicenter*. In the previous chapter I discussed Robert Webster and Graeme Laver's intriguing, but largely disappointing, 1972 trip to China. Conferences and exchanges of virus samples increased during the 1980s and 1990s, but international scientific research on pandemic influenza remained at a remove from the agricultural landscapes of the epicenter. China's laboratory infrastructure was slow to develop, which impacted domestic research and the international sharing of samples. As Guo Yuanji of China's Institute of Virology explained in a 1985 letter, China had not sent "original specimens . . . to the WHO Collaborating Center for reference and research on influenza in London or Atlanta for final identification" because the Institute lacked "a good quality refrigerator to keep speciments and reagents."[44] And as virologist Gavin Smith recalls, despite the success of the Hong Kong laboratory's experimental system in sampling and typing viruses, it was "too difficult politically and too expensive" to get to the source of the animals on farms inside mainland China.[45]

As the interdisciplinarity of the Bangkok meeting hints, investigating the ecology of viral emergence opened research on pandemic influenza to many scientific perspectives, including remote sensing of landscapes from satellite imagery, tracking of wild-bird migration, surveys of poultry density, and modeling species interactions. Yet as Marius Gilbert later told me, these new perspectives typically required investigation or validation "on the ground," inside the actual landscapes of the epicenter. As a result, researchers like Gilbert, Xiao, and others began to move their experimental systems from laboratory venues to field settings, and from scientific centers, including Geneva (WHO) and Rome (FAO), toward the remote sites identified as hypothetical epicenters inside China.

In the rest of this book I explore the consequences of these displacements on the morphology of global health and scientific research on pandemic influenza. To exit the laboratory means not only to leave behind the microscopes and virus cultures that organize the epistemic and experimental system of viral phylogenetics. In anthropology and science studies, the laboratory is much more than the iconic, white-walled house of experiment; more importantly, the lab also invokes a distinctive model of scientific authority and expertise. To exit the laboratory, therefore, is also to raise new questions about the proper relationship between scientific knowledge and other modes of life, production, and politics. In moving from the laboratory to the epicenter, the validation of a new epistemology of ecology required unexpected engagements with the cultural and political dimensions of China's landscapes.

PART II
LANDSCAPE

CHAPTER THREE
LIVESTOCK REVOLUTIONS

The lowland region surrounding the Poyang Lake is crisscrossed with an intricate system of man-made waterworks. High, earthwork dikes form rings around the lakeshore and line the riverbanks to prevent floods and extend agriculture onto fertile riverine soil. Irrigation ditches run alongside most roads. Small ponds, initially dug for fish farming, are scattered around villages. Today, many of these water bodies have been repurposed for duck farming and are populated with flocks of swimming white birds. Duck sheds, made of wooden slats and plastic sheeting, often stand at the water's edge (see figure 3.1). "To drive anywhere," wild bird researcher Scott Newman once remarked with amazement about the lake, "from *any* point A to point B, you see lots of duck farming."

The scale of duck farming around the Poyang Lake is a local consequence of the livestock revolution that accompanied China's post-Mao market reforms. From 1970 to 2017, China's annual meat chicken production (per head) has grown from around 600 million to almost 10 billion, but duck production, though beginning from a smaller starting point, has grown just as fast. Duck production increased around fifteenfold, from around 150 million (1970) to 2.25 billion (2017).[1] China now accounts for roughly three-fourths of ducks produced in the entire world (see figure I.2).[2] In the early 1980s, Jiangxi Province planning reports called for Poyang Lake to be developed as a "production base" for commercial waterfowl. According to recent data collected from county-level

FIGURE 3.1. Adjacent duck sheds on canal.

agricultural yearbooks, there are more than 14 million ducks raised in the region today (around half as many as in the entire United States).[3]

The high visibility of ducks—easily seen from a van speeding along roads and highways, as Newman pointed out—also reflects the mode of husbandry common in the region. Most ducks are raised at least partly outdoors in what farmers refer to as *fangyang* or "free-grazing." In some instances, farmers let the ducks wander free; in others, they bring the ducks out of sheds and enclosures to graze under supervision in fields, canals, and ponds. Rice paddies, in particular, play an important role. Paddy-rice cultivation makes up over 90 percent of the sown area of all crops in the Poyang Lake region.[4] In an agricultural technique known as *daotian yangya*, or "rice-duck coculture," that dates back centuries, farmers purposefully bring ducks to graze in rice paddies. In this way, ducks get supplemental feed by gleaning rice left after harvest and eating weeds or small insects. At the same time, they provide the rice crop with services such as pest control and fertilization.[5]

As I showed in the previous chapter, scientific research on pandemic influenza hypothesizes that a pathogenic landscape lurks amid this pastoral scene. Indeed, Kennedy Shortridge's discovery of a reservoir of influenza viruses in domestic ducks directly influenced his identification of southern China as the "epicenter" of influenza pandemics. In their classic 1982 article on the influenza epicenter, Shortridge and C. H. Stuart-Harris wrote that rice-duck coculture, in particular, "provides a close bird/water/rice/man association that varies with the seasons of rice-growing" and thereby promotes the "interchange of viruses between host species."[6] Recently, Marius Gilbert and colleagues showed that the spatial distribution of H5N1 influenza outbreaks closely matched the distribution of "traditional free-grazing duck husbandry."[7]

In response to the pandemic threat posed by avian influenza, these ecological diagrams of viral emergence have been mobilized in two interconnected ways.[8] First, these diagrams enable particular places, such as Poyang Lake, to be identified as areas of relatively high risk for HPAI outbreaks or viral emergence. As one Chinese-led study notes, the Poyang Lake area is a zone of increased "disease risk" because "most backyard poultry raised in the Poyang lake area are in a free ranging style."[9] One graduate student I met wrote a master's thesis at a local Jiangxi university that compared known risk factors such as "free-grazing poultry"—"drawn from expert knowledge and experience" that the student collected from published papers—to the actual conditions and farming practices in the Poyang Lake region. One might say he superimposed the zoonotic diagram onto the forms of the actual landscape. Second, ecological diagrams of viral emergence, and particularly those that trace a chain of interspecies infections, also mark out points of intervention, sites of prevention, or targets for disrupting chains of infection.[10] In this regard, diagrams highlighting the role of free-grazing ducks focus attention not only on a particular species but also on practices such as husbandry techniques (free-grazing) or farm infrastructure (enclosed housing).

This chapter examines how this scientific model of viral emergence became the basis for a program of biosecurity interventions on China's farms in the global quest to prevent the next pandemic. I trace the Food and Agriculture Organization's initial response to the reemergence of HPAI H5N1 in 2003, and in particular how the FAO translated pandemic preparedness into the idiom of rural development through the concept of *biosecurity*. Whereas WHO programs for pandemic preparedness largely focused on national response planning, virus surveillance, and the development of new vaccines, FAO officers argued that the way to prevent a pandemic was to control the virus "at source." In a plea for donor funding, the FAO explained that "The Avian Influenza (AI) virus has had a considerable economical and social impact on affected countries and could potentially lead to a new global human influenza pandemic. Therefore, it is in the interest of developed and developing countries to invest in the control and containment of the virus. Control of highly pathogenic avian influenza

(HPAI) at source means managing transmission of the virus where the disease occurs—in poultry, specifically free range chickens and in wetland dwelling ducks—and curbing HPAI occurrence in . . . Asia before other regions of the globe are affected."[11] For the FAO, "targeting the disease <u>at source of infection</u>" meant intervening in two domains: geographically, in Asia, and ecologically, in free-grazing poultry.[12]

The FAO justified its own involvement in part according to its traditional mission to protect livestock-based economies and food security. As Phil Harris, an FAO information officer, put it, "Avian influenza remains a potential risk to humans but a real risk to animals." At the same time, the FAO for the first time began to position its work as a crucial component of pandemic preparedness and bulwark for human health. In this regard, Harris pointed out that "where animal disease poses a potential threat to human health, FAO's role is to advise on the best methods to contain the disease at the *level of animals*, prevent its recurrence and undertake research to identify ways of eradicating the disease. . . . The current state of play is that avian influenza is an animal health issue and the focus must be on attacking the problem at source—in animals."[13]

Pandemic preparedness operates in a mode of *potential* uncertainty—that is, pandemics are thought about and acted upon as incalculable futures and therefore uninsurable risks.[14] Although the probability that a particular strain will cause a pandemic cannot be statistically enumerated, the potential threat of a pandemic would cause catastrophic consequences for global health. Indeed, avian influenza viruses such as the H5N1 strain are often referred to as viruses with "pandemic potential."[15] Social science scholarship on pandemic preparedness has drawn attention to programs such as vaccine development and stockpiling, the construction of early-warning surveillance systems, and scenario-based exercises—emphasizing how these techniques aim not to prevent an emergency but to survive or mitigate one.[16] By contrast, the FAO's work in Asia shows that the anticipation of the potential pandemic is configured differently in the spaces defined as "sources" of pathogens than in the wealthy countries that seek "self-protection" from pandemic threats.[17] The potential uncertainty of the next pandemic brings global health from labs and hospitals to the rural farms of the hypothetical epicenter. And on the farm, even if a pandemic threat is only a potential future, the logic of emergency drives immediate—and sometimes violent—interventions into the bodies and relations of animals and humans.

In this chapter I examine how this mode of uncertainty shaped interventions into China's hypothetical influenza epicenter—including culling of diseased birds, closing of markets, and long-term efforts to reduce the number of smallholder poultry farms and free-grazing practices. I track the development of FAO policies for pandemic preparedness alongside China's own biosecurity planning. Then, drawing on my fieldwork with poultry farmers in the Poyang Lake region, I show how biosecurity programs intersected with China's livestock revolution in unexpected ways. In particular, I compare the FAO's assessment of free-grazing poultry disease risk with the very different risk assessments made by poultry farmers in the Poyang Lake region.

To be clear, I must distinguish this from the comparison of expert and lay risk perceptions that characterizes research into the public understanding of science.[18] Rather than seeking to uncover the local knowledge, implicit distrust, or critical potential of poultry farmers, I instead highlight the interaction between two anticipatory states: (1) the *potential uncertainty* through which flu scientists turned southern China (and Poyang Lake) into a hypothetical source of future pandemics, that is, a pandemic epicenter, and (2) the *vital uncertainty* that surrounds the work of duck farming at the Poyang Lake, where duck farmers configure market regimes of debt and profit with the growth and life of the ducks. By documenting this interaction, the chapter shows how the initial encounters at the pandemic epicenter between global health and agricultural production increased, rather than reduced, uncertainty for both farmers and global health models. The unexpected dimensions and scales of Poyang Lake's working landscapes displaced efforts to impose control "at source."

TEMPOS OF PREPAREDNESS

The FAO's biosecurity interventions moved into the pandemic epicenter according to three distinct tempos: an emergency response to an urgent outbreak (seeking *eradication* of a circulating virus), an immunological approach to containment, and a developmental program to reform agricultural practices in order to "reduce risks of outbreaks as a long-term measure."[19]

At first, in response to outbreaks of highly pathogenic avian influenza viruses, the FAO recommended the culling of infected and surrounding flocks of poultry and the institution of movement and trade controls, or what is known as a policy of "stamping out": "When outbreaks of H5N1 or other HPAI viruses occur, immediate stamping out is the most appropriate and the first response of Veterinary Authorities and it is most likely to be successful when it is combined with movement controls, decontamination of infected premises and proper surveillance and monitoring."[20]

Until the mid-1990s, culling birds on and around affected farms had successfully controlled all HPAI outbreaks. Most of these outbreaks took place in the industrial poultry sector in Europe or North America, with the largest out-break striking Pennsylvania farms in 1983–84. Only during the 1990s, during HPAI outbreaks in Mexico and Pakistan, did culling begin to show its limits. In a 2000 review article, leading avian influenza researchers David Swayne and David Suarez still stated that "eradication is the only viable option" for HPAI, discouraged strategies based on control or management of disease, and recommended a "stamping-out or slaughter program for all HPAI" outbreaks.[21] Abigail Woods, in a history of foot-and-mouth disease in the UK, has pointed out that the urgency of "stamping out" often exceeded scientific or epidemio-logical utility, leading to costly slaughter of animals without effective control of the disease.[22] It is certainly true that the specter of an imminent human pandemic heightened the urgency of poultry disease control at the epicenter, driving the slaughter of poultry in Asia to an enormous scale. More than 100 million chickens were culled in Asia between 2004 and 2005 alone.[23]

Culling was usually combined with movement and trade controls, such as closing of live-poultry markets. Although such control measures appear less violent than forced slaughter, the impact on farmers was often enormous and stretched far beyond the outbreak zone. Because compensation programs for culled birds or lost markets were usually insufficient or nonexistent, farmers lost capital investments and sometimes even subverted the objective of bio-security interventions by illegally selling poultry to traders at cut-rate prices. Within China, a debate emerged about appropriate compensation levels and the efficacy of market closure.[24]

Second, FAO officials and others raised concerns that "stamping out" was not eradicating the virus from farms in China or several other Southeast Asian countries (especially Vietnam and Indonesia). Indeed, HPAI H5N1 increasingly appeared to be an endemic, rather than epidemic, disease. At a conference in Chiang Mai in September–October 2004, Senior Officer Juan Lubroth (FAO, Emergency Prevention System for Animal Health [EMPRES]) noted that "while stamping out is the preferred option vaccination is a suitable tool and can be used as a precursor to stamping out." Lubroth suggested that stamping out was most effective in situations of "early detection/small outbreak," whereas "vac-cination as an ancillary tool" was suitable in cases of "delayed detection" or where there were "limited abilities" to achieve eradication. By late 2005, China would embark on a universal mandatory vaccination program.[25]

Third, the FAO also sought to transform the ecology and epidemiology of the epicenter over a longer duration.[26] To ensure eradication and the more permanent transformation of the epicenter, FAO position papers called for changes to the ways farmers in rural China farmed poultry. At the pandemic epicenter, as research moved from viruses to the agro-ecologies that drive viral emergence, the apocalyptic framing of future emergency became entwined with a more long-standing orientation to the future: rural development. In what I call a model of *biosecurity development*, the FAO proposed the "restructuring" of duck farming as a long-term approach to pandemic risk. This model linked improvements in farm biosecurity with increases in scale and commercial integration, suggesting that a linear trajectory of agricultural development would enhance pandemic preparedness.

In a series of position papers on the improvement of farm "biosecurity," the FAO charted a new future for the poultry farms inside the influenza epicenter.[27] What did the FAO mean by this highly polysemous term?[28] Biosecurity for avian influenza involves the "implementation of practices that create barriers in order to reduce the risk of the introduction and spread of disease agents," according to a 2008 technical paper.[29] The three principal elements of biosecurity are segregation, cleaning, and disinfection: each is a technique of separation.[30] Biosecurity regulates the conduct of relations between living beings, including those between humans and animals, between poultry of different species, and between domestic and wild animals.[31] Biosecurity practices range across a wide variety of scales. Gloves or plastic shoe covers help to maintain barriers between humans and animals. The construction of enclosed housing keeps domestic birds separated from wild birds. Disinfection protocols, including the targeted use of disinfectant sprays, keep each shed or farm separated from others.

The poultry shed, in which the flock is enclosed and protected, is the icon of biosecurity. One FAO report noted that "a well-organized entrance with a barrier, used to exclude most people and objects," makes for the "single most important measure that any poultry unit can take to decrease the risk of infection."[32] Conversely, the free-grazing flock is figured as the greatest threat or obstacle to biosecure poultry farming. "While it is feasible to tighten biosecurity on commercial poultry farms," an early paper points out, "this may be more difficult, or impossible, in the case of non-commercial enterprises, such as back-yard production systems, particularly where flocks forage outdoors."[33]

In the initial response to the avian flu crisis, the FAO situated the pathway of biosecurity improvement, including material changes such as housing construction, within a broader model of economic transition. The route toward biosecurity, according to several FAO papers, lay in the "restructuring" of the poultry sector: "Restructuring of poultry sectors will play an important part of long term prevention. . . . Successful restructuring will play a critical role improving biosecurity to prevent spread during the current epidemic and reduce risk of outbreaks as a long-term measure."[34] Through this model of restructuring, pandemic preparedness is synchronized and overlapped with developmental models of economic change. Biosecurity at the epicenter brings emergency and development into a common space and time.

Indeed, much like classical modernization and development theories, plans for restructuring rely on a step-based process of transition and a typology of developmental states. What is new is that in the analytic of restructuring, the developmental process is now paired with improvements in control or prevention of disease. Each developmental stage is not only an economic transformation but also a reduction of biological risk. In one of the organization's first position papers on the avian influenza crisis, the FAO developed a conceptual distinction between four different "production sectors" of poultry farming. The sectors are differentiated according to two factors: level of "farm biosecurity" and "system used to market products." Yet these are not treated as two independent variables. Rather, level of farm biosecurity is tied to marketing system in the definitions of each model sector:

SECTOR 1 Industrial integrated system with high-level biosecurity and birds/products marketed commercially (e.g., farms that are part of an integrated broiler production enterprise with clearly defined and implemented standard operating procedures for biosecurity).

SECTOR 2 Commercial poultry production system with moderate to high biosecurity and birds/products usually marketed commercially (e.g., farms with birds kept indoors continuously; strictly preventing contact with other poultry or wildlife).

SECTOR 3 Commercial poultry production system with low to minimal biosecurity and birds/products usually entering live-bird markets (e.g., a caged layer farm with birds in open sheds, a farm with poultry spending time outside the shed, a farm producing chickens and waterfowl).

SECTOR 4 Village or backyard production with minimal biosecurity and birds/products consumed locally.[35]

The typology implies that increases in farm scale and industrial integration will inherently bring improvements to biosecurity. As Olivier Charnoz and Paul Forster have asked about this typology, "The question here is how could anyone who subscribes to this world-view not see the lower numbered sections as being more implicated in the generation and spread of a disease?"[36]

In a 2007 paper, Olaf Thieme and Emmanuelle Guerne Bleich of the FAO's Animal Production Service define "restructuring" as "change of a production system through external interventions," in contrast with structural change "driven by market forces." The primary purpose of restructuring is "to increase bio security in order to help with control & eradication of disease and make products safer for consumers." Indeed, Thieme and Guerne Bleich emphasize that the "main purpose [of restructuring] is to make production safer not to increase size of operations." However, in a "plan" developed for a case study in Vietnam, they outline three concrete interventions: (1) increase proportion of big commercial farms; (2) relocate commercial farms to specific production zones, out of urban areas; and (3) redirect market chains toward selling of processed poultry.[37] They note that the impact of this plan on farmers, especially duck farmers, would be "substantial" and that in the case of duck farming in particular, "planned restructuring will lead to a completely new production system."

Instead of suggesting a simple trajectory of increased scale, Thieme and Guerne Bleich are proposing a process of integration and standardization that is closely associated with larger-scale, capital-intensive poultry farming. Critical geographers have documented the ways in which biosecurity and food-safety regulation are linked to, or promote the development of, capital-intensive and industrialized forms of agricultural economy. John Law, for example, notes that the dominant response to the United Kingdom's foot-and-mouth disease (FMD) disaster reflects a hope for "uniformity" in regulation of meat production. "The aspiration," he writes of contemporary industrial agriculture, "is to standardize flows and exchanges on a global scale."[38] Hinchliffe and colleagues have further highlighted how the "purifying schemes" and "will to closure" of farm biosecurity often support particular business models of large-scale, vertical integration. "In the industry," they write, "biosecurity is perceived to be more effective in tightly coupled, highly integrated production processes, where large organizations can effectively exercise control across the length and breadth of the food chain and design-in barriers and buffers to keep the system disease-free."[39] More broadly, Elizabeth Dunn has shown how food standards favor large-scale industrial farms and often force consolidation. Writing of the introduction of European Union food-safety standards to Poland, she shows that the strictness of standards has the effect of driving out small slaughterhouses that cannot keep up with capital-intensive upgrades needed to meet regulatory requirements.[40] And other critics have suggested that the infrastructures and methods of intensive, industrial poultry farming in fact play a significant role in the emergence and transmission of avian influenza viruses, precisely due to their high throughput, tight coupling, homogenous breeds, and large-scale trading networks.[41]

When pandemic-preparedness programs first moved into the epicenter, FAO plans called for the restructuring of the poultry sector. The *potential* uncertainty that a human, global pandemic will emerge from farm ecologies in southern China drove an intensifying effort to enclose and separate relations among animals and between species. Further, these plans proposed that biosecurity interventions—such as construction of enclosed housing—should be linked with economic transitions toward large-scale, integrated farming. They synchronized and overlapped two distinct logics of space and time: global emergency (the potential uncertainty of a future pandemic) and rural development (a linear transition from "backyard" production to "industrial integration"). Restructuring rural China's agriculture, they proposed, would help prevent the next global pandemic.

SCALING UP

During my time working with the Emergency Center in Beijing, I found that China's Veterinary Bureau readily adopted the FAO's biosecurity discourse, and even more readily its focus on developmental transformation of small-scale and free-grazing farms. Indeed, the officials and state-employed veterinarians that I met tended to elide the FAO's four sectors into a simpler opposition between "scale" (*guimo*) and "dispersed" (*sanyang*) farms. When developing a mock farm survey during a training program that I attended, for instance, the state-employed vets spent much of their time identifying a suitable quantitative boundary to distinguish "scale farms" (*guimo chang*) from "scattered" or "household farms" (*sanhu, nonghu*), eventually selecting two hundred for pigs and one thousand for poultry. They also explained to me that China's poultry industry was distinctive, compared with countries like the United States, because of the concurrence and adjacency of scale and scattered farms. For example, one young veterinarian told me that a particular area of Henan Province was suitable for a research project because it was "representative": "these areas have *both* scale husbandry [*guimo yangzhi*] and scattered [*sanyang*]," he explained.

This typology can be (and often is) loosely correlated with the commonplace English-language classifications of farms as "industrial" and "backyard," but there are particular meanings associated with the Chinese terminology. First, *guimo* ("scale") is used to mean *large* scale. The term *scale* is shifted in this usage from a measurement tool to a relational position on the ruler itself.[42] The implication is that *sanyang* poultry farms not merely are small scale but, more importantly, have not crossed a critical threshold to *achieve* scale.

Second, by opposing "scale" to "scattered"—rather than merely opposing large and small scales—the terminological distinction equates the scale of farm size with the form of husbandry practice. *Sanyang* (literally, "scattered husbandry") means to raise livestock or poultry outside of a pen or enclosure: to free graze. In this sense, "scale" implicitly refers not only to the size of farm or number of animals but also to the built structures and farming techniques through which animals are enclosed and managed. The opposition draws on a model of agricultural modernization based in the European and American historical experience, in which poultry were moved indoors and raised under increasingly controlled conditions as the scale of farms grew.[43] It also suggests that large-scale farms, associated with infrastructures of enclosure and containment, are more biosecure and less vulnerable to infection.

State-employed veterinarians referred to the objective of agrarian development during the post-Mao era as a trajectory of *guimohua*—a process of "reaching scale" or "scaling up." Andrew Donaldson and Qian Forrest Zhang helpfully distinguish the reform era into two moments of agrarian change. In the first, during the late 1970s and 1980s, the state divided communal farms and distributed land-use rights, as well as production and supply quotas, to households: the so-called Household Responsibility System. Farm sizes were very small, averaging 0.7 hectares per household.[44] At the same time, rural markets opened, and household farmers began to undertake sideline—and then specialized—farming of cash crops or livestock for market sale. By 1990, however, Deng Xiaoping was calling for a second stage of agricultural reform, or what Deng called a second "leap": "The reform and development of China's socialist agriculture, from the long-term perspective, requires two great leaps (*liangge feiyue*). The first leap is dismantling peoples' communes and implementing the Household Responsibility reform. This is a great advance and should be kept in the long term. The second leap is meeting the needs of scientific agriculture and socialized production, properly developing scaled-up operation, and developing the collective economy."[45]

The second leap, as Deng elaborated, involves mechanisms such as *collectivization* and *intensification* in order to "industry-ize" (*chanyehua*) rural agriculture—that is, to help transition rural farming from a primary mode of production to secondary production that would process agriculture goods into products. The goal of government policy in this area is to replace "one family, one household scattered [*fensan*] production" with organizational models that will "intensify" land use, increase the scale of agricultural production, and reduce risk for farmers.

Crucially, however, intensification and increased scale would not replace household-scale farms as the "predominant production unit" but would rather aim to transform them through what Phillip C. C. Huang has called "new-age small farming."[46] China's agricultural development policies aimed to achieve economies of scale through mechanisms of vertical integration. According to Huang, the concept of vertical integration "gained the stature of official sponsorship" and was frequently used "to represent the present and future of China's agricultural modernizations."[47] Although there are several mechanisms of vertical integration, all aim to unify a large number of small-scale primary producers within an integrated, or centrally managed, supply chain from production through processing to marketing.

By far the most significant of these mechanisms of vertical integration has been for-profit agribusiness "dragon-head enterprises" (*longtou qiye*).[48] Dragon heads are private agribusinesses that are designated by central and local governments for their role in vertically leading farmers into processing and marketing through the formation of integrated supply chains. By meeting certain criteria in terms of size and financial profile, enterprises designated as dragon heads are provided with subsidies, preferential tax treatment, and loans.[49] Some of the most successful dragon heads are in the poultry sector, particularly in broiler farming, including Thailand's CP group and Guangdong Province–based Wen's Foodstuff. Initiated in 2000 with 151 enterprises, there were 110,000 officially designated dragon heads by 2011. Moreover, at least according to official statistics, these dragon heads had enrolled 110 million rural households as producers and accounted for more than 60 percent of crop production area, 70 percent of livestock (pigs and poultry), and 80 percent of aquaculture production.[50]

In the wake of the pandemic influenza threat, the association of emerging disease risks with backyard poultry provided a new justification for scaling up poultry farming. Figuring rural China as suffering from an incomplete transition to scale, veterinarians argued that food safety and animal disease control faced unique challenges in contemporary China. "There are too many small farmers," a dean from the China Animal Health and Epidemiology Center explained at a conference on food safety in animal products I attended in 2011. "The level of intensive farming is too low." As the director of China's Veterinary Bureau elaborated in a published interview, China's livestock husbandry is still predominantly "scattered," with only a small proportion of scale (*guimo*) farms, rendering state surveillance and management extremely difficult.[51] This opposition between large-scale (biosecure) and small-scale (low biosecurity) was also applied at Poyang Lake to explain the distinctive pandemic risk of the region. In a report on risk factors at Poyang Lake, a Chinese team conveys what they claim to be a description made by a "local veterinarian": "Local veterinarian told us that commercial farms usually operate on a large scale and poultry is generally kept in high density and enclosed housing conditions with high biosecurity measures. On the contrary, the majority of backyard poultry owners raises the poultry in open yards in a small scale, and do not apply much biosecurity measures. The flocks are raised for owners' consumption only."[52]

Was China's livestock revolution following the pathways toward biosecurity development and increased scale outlined by the FAO and the Ministry of Agriculture? How was quantitative scale configured with qualitative mode of husbandry?

QUANTITIES AND QUALITIES

At Poyang Lake, I found that scale doesn't always increase according to plan. As I traveled across Nanchang, Poyang, and Yongxiu counties, which surround the south side of the Poyang Lake, I saw that scale was often an emergent property of a landscape. Alongside the long irrigation ditches that line the roads, duck sheds often perch on the artificial embankments just above. In other areas, duck sheds belonging to different households are located adjacently, sometimes gathered in groups around a common pond, forming informal "husbandry zones" (*yangzhi xiaoqu*) that mirror the cluster of houses in the villages. Although each household might raise only two or three thousand ducks, these emergent, loosely coordinated agglomerations can reach scales of hundreds of thousands of birds (figure I.2).

Confronted with the actual landscapes of the livestock revolution, it was clear to me that the FAO's typology of sectors did not capture the qualitative changes in poultry farming. Searching for a more anthropological analysis of these changes, I found sociologist Li Huaiyan's description of transition from "sideline" to "specialized" modes of animal husbandry helpful. Drawing on a long-term study of a Jiangsu village, Li traced how the distribution of collective farmland to households drove a social transformation of livestock farming practices. Li points out that before the 1980s, most farm labor was devoted to work on collective fields and this work was paid for in shares of the collectively harvested crop. At the same time, however, "each household raised a limited number of chickens, pigs, or goats only to augment its primary income from the production team." Because this was secondary work, done in spare time, "they did not care much about . . . cost and profit." Li calls this mode of farming the *sideline.*[53] As historian Jonathan Unger has noted, the Maoist state generally allowed sideline farming, but certain political campaigns, such as during the Cultural Revolution, "forced through reductions in the size of private plots [and] implemented very strict limits on the number of ducks and chickens farmers could raise."[54]

After the 1980s, in Li's historical account, and especially during the 1990s, "Raising a large number of such animals became a family business that generated most of the family's income. Therefore, the family had to carefully calculate its labor and capital input to make a profit." Li calls these "specialized" households (*zhuanyehu*), adopting a term used in rural development policy programs during the period to indicate a household that focuses on the production of a single crop or livestock breed.

Li's diagnostic distinction is crucial, and it shows how shifts in farm scale were tied to shifts in mode of production and farmer subjectivity, but its historical narrative is too linear. His account implies a unified historical transition from sideline to specialized farm that closely resembles the developmental model from *scattered* to *scale*. In the Poyang countryside, I found that although the geography of duck farming does contain an opposition between sideline and specialized, the opposition is not structured as a developmental transition. For by and large, I found, those farmers who have begun to "specialize" in raising ducks as part of a commercial business *also* continue to raise another flock of ducks "on the side."

Consider Chen, for example, a duck farmer I visited whose house stands beside the main road near the Lian River, south of Nanchang. Chen invited us into the front door of his house and then immediately showed us out the back door, where a poured concrete courtyard sloped down into a man-made pond. Around forty ducks and geese swam in the pond, ranging across a wide variety of sizes, colors, and breeds. Pointing at the birds, I began to ask some of the questions I had prepared. Did the household primarily earn money from selling eggs or from selling these birds for meat?

Chen stared at us for a moment, as if not comprehending the question, and then burst into laughter. "These birds? These birds aren't any good to sell! All of these our family raises to eat, for eating at the New Year!"

As we laughed together at my mistake, his wife suddenly pointed across the rice fields: "But we do raise those other ducks down there. . . ." Now I was the one who was confused. Chen kindly offered to show us, and brought us to a shed about a ten-minute walk across the rice fields. Here we saw five thousand layer ducks, raised by the family for eggs that they sold at the Xiaolan wholesale market. Eventually, when laying productivity began to decrease, they would also sell the ducks to a broker who would bring them to a live-poultry market for sale as meat. Here was a surprising scale boundary, a distinction of quantity and quality drawn within a single household. Chen raised two flocks of ducks and separated them physically and conceptually: one flock for the family to eat and one flock for the market.

I soon found similar practices at farms across the Poyang Lake region. There are three modes of separation that distinguish the two flocks. The first is the physical relationship of the birds to the family home. As we saw, the Chen family keeps ducks and geese raised for family consumption immediately behind the house. The birds freely roam around the small pond there and are allowed into the rice fields that surround the house on three sides. Notably, although plastic fencing surrounds the small kitchen garden to keep the ducks *out*, there is no fencing to keep ducks *in* anywhere. In some of the households I visited, small sheds are built for ducks to spend the night. In many cases ducks are left in the open air. Farmers assured me that ducks huddle together for warmth and will not suffer during the cold winter. During the day, ducks move freely. Ducks are famous for their ability to find their own way home at night: in one village I stayed in, the ducks returned from the river or rice fields in the evening, peeling off in twos or threes to return to the households that raise them.

By contrast, the Chen family keeps ducks raised for the commercial market at some distance from the house, where they are enclosed by wire mesh fencing and a small duck shed (see figure 3.2). Duck sheds are often built of wooden or bamboo poles and partly wrapped with plastic sheeting. Others are more durably constructed out of brick or concrete blocks. They are erected next to small man-made ponds, canals, and other bodies of water. A fence encircles a stretch of land in front of the shed and a portion of the water for the ducks to swim in.

The redistribution of land usufruct rights from collectives to individual households was a precondition for the emergence of what Li Huaiyan called the "specialized" farm household. However, the growth of specialized households engaged in farming of ducks for the commercial market has given rise to additional transformations of village space. For example, in Shibahu, the village I stayed in when I visited the Wang family's wild goose farm (to be discussed in chapter 4), the commercial duck sheds are separated from each household, but they are also located close to the duck sheds of other households. Sometimes this spatial arrangement is formalized in village policy. According to law, rural land is divided into three categories: farmland, construction land (i.e., for household construction), and unused land. Crucially, "basic farmland" cannot be converted from crop cultivation to other uses, such as animal husbandry. In

FIGURE 3.2. A typical duck shed near Poyang Lake.

response to the growth in commercial livestock farming, the 2006 Livestock Law directs local governments to make comprehensive land-use plans that designate some land for crop cultivation and other land for livestock farming, referred to as "husbandry zones" (*yangzhi xiaoqu*).[55] Villagers are able to contract land in these zones for raising poultry and even constructing permanent sheds, but after the contracts expire, land must be returned to its original use.

Two distinct spatial patterns emerge: the ducks raised for the market are both *separated* from the space of the family's house and partially *enclosed* by physical structures and fencing. By contrast, the ducks raised for family consumption are *adjacent* to the family household and allowed to roam freely without enclosure.

Feeding is a second axis of separation. The Chen family feeds the ducks that they raise for their own consumption leftover scraps from the kitchen table, along with unshucked and unpolished rice harvested from the household's own rice fields. Chen emphasized that he fed the ducks in the pond by his house *exactly* the same foods eaten by the family. By contrast, ducks raised for sale in the market are primarily fed manufactured feeds. In the Poyang Lake region, farmers often buy feed in large sacks at the wholesale egg market where they deliver their eggs. When Chen took us to see the ducks in the large shed, he waved around a bag of feed and scattered some grains to attract the birds toward us. By contrast, he reiterated that the ducks he farmed for his own family's consumption "never eat manufactured feed."

Practices of eating together, or commensality, are fundamental to the formation of social relationships, as well as the marking of social boundaries through inclusion or exclusion.[56] As Marilyn Strathern has recently pointed out, however, *feeding* practices can produce or break social relations just as powerfully as *eating* practices can.[57] In this sense the feeding of leftover scraps to the family ducks enacts a kind of trans-species commensality. Although the ducks are not eating at the same table as the family, or literally "breaking bread" together, they are eating the same foods cooked in the same pots. The animal pet provides an insightful model for thinking here. In a classic essay on human–animal relations, Edmund Leach suggested that English attitudes of friendly familiarity toward pigs were partly shaped by the (historical) status of pigs as "commensal associates": "pigs, like dogs, were [until recently in rural England] fed from the leftovers of their human masters' kitchens."[58] Fausto and Costa have recently drawn attention to Amazonian practices of feeding pets as productive of asymmetric, rather than commensal, social relations. In some cases, though, when the bond between an owner and her pet becomes strong, the "relation of feeding veer[s] towards commensality."[59]

In the Poyang Lake region, the feeding of leftovers and household rice to ducks is a semi-commensal practice that increases nearness between ducks and their masters. Conversely, the feeding of commercially manufactured feeds excludes the market-bound ducks from commensal relations with the family. This is particularly significant because the rise and growth of commercial duck farming are closely, even intrinsically, tied to the growth of the processed animal feed industry. Manufactured feeds enable the independence of duck farming from the constraints of natural contexts and local settings, and therefore remove limitations on the quantity of ducks raised in a particular space.[60] As inputs of feed grow, so can outputs of poultry eggs or meat in an almost linear fashion, no longer limited by the quantity of available surplus grain or foraging opportunities in the surrounding landscape. Ke Bingsheng, a scholar of China's rural economy, argues that "the emergence and development of the processed feed industry have played a decisive role in shaping the structure of the livestock sector in China." Indeed, the growth of the feed industry, which Ke points out "developed literally from scratch" in the late 1970s, directly parallels the growth in poultry production.[61] Yet ducks raised for family consumption are never fed commercially purchased feeds. Instead, they are fed leftovers and household rice. Through feeding practices, the ducks raised for household consumption are "brought close" to the household, while ducks raised for market sale are "pushed farther" from the household.[62]

A third mode of separation is reflected in the *purpose* and outcome of production: the ducks raised in the sheds are produced for the market, whereas the ducks raised behind the house are destined for family consumption. In order to further develop this important distinction, however, we need to look more closely at the particular form and trajectory of the layer-duck market in the Poyang Lake region.

CENTURY EGGS AND OLD MOTHER DUCKS

Most commercial duck farmers in the Poyang Lake region raise *layer* ducks, that is, egg-laying ducks. Approximately every week or two, farmers join together with kin or village associates to rent trucks and deliver eggs to the regional Xiaolan wholesale egg market just south of Nanchang. The Xiaolan market is one of the largest wholesale egg markets in China, and farmers from across the Poyang region sell eggs there. When the Xiaolan market opens around six in the morning, the buyers of duck and quail eggs who lease shop space begin to set up their scales. Soon after, farmers pull into the marketplace in large trucks. In most cases, four or five farmers engage a driver and a truck to bring their eggs to market. The farmers sit with the driver up front in the cab, with wooden boxes of eggs piled high on the truck bed. The trucks slowly lurch to a stop, unloading their passengers near the center of the market. At this central square, farmers and buyers meet in small agonistic clusters to haggle the daily price of eggs. Beneath a high plastic and metal roof, open to the air at one end, the egg exchange is dim, lit only by indirect, raking sunlight (see figure 3.3).

At the Xiaolan market the dominant buyers of duck eggs are the representatives from local *pidan* ("century egg") factories. These factories preserve duck eggs in a mixture of clay, ash, salt, quicklime, and rice husks until they become almost black. Even more than the initial market transaction, the preservation process crystallizes value, confirming the transformation of duck eggs into commodities. On the one hand, a rather-poor-tasting duck egg (in China, rarely eaten fresh) is turned into a delicacy. On the other, a fresh egg with a limited shelf life becomes a preserved egg, enabling the factories to ship century eggs to consumers across China and even export them internationally.

For these reasons, it is not surprising that the preserved-egg factory bosses wield asymmetric power in the determination of prices negotiated at the wholesale market. Even in the dim light, the factory owners and reps clearly

FIGURE 3.3. Trucks inside the Xiaolan Poultry Wholesale Market south of Nanchang.

stand out from the farmers, several driving shiny black luxury sedans into the dusty marketplace. Even though I could see that farmers were probably getting a raw deal and had few alternative outlets, I was still confused when I was first told that farmers make no money at all from the sale of eggs. Although prices fluctuate, on most days farmers only break even once they factor in the purchase cost of manufactured feeds. They purchase feeds in large sacks, from stalls located in the same marketplace, and load the sacks onto their trucks to bring back to the village. By watching the trucks entering and leaving the market, one might mistake the economy for a bartered exchange of eggs for feeds (see figure 3.4). And indeed, some farmers told me they are not even paid in cash, but rather in credit slips that must be used to directly acquire feeds from vendors at the market. These farmers suspected that preserved-egg factory bosses colluded to hold the price of eggs down or conspired with feed vendors to ensure profits.

Gradually, I learned that whether or not the preserved-egg bosses actually fixed prices, the sale of eggs was never imagined to be the only way farmers make money from raising layer ducks. Indeed, the farmers earn much of their money when the egg-laying productivity of the ducks begins to decline at the end of a one-year cycle. These *spent ducks*, as they are called in English, or *lao-*

FIGURE 3.4. Groups of farmers rent trucks to transport eggs to the Xiaolan wholesale market. A truck loaded with manufactured feeds and the now-empty crates used to transport eggs is prepared for the journey home.

muya ("Old Mother Ducks") as they are more melodiously named in Chinese, are a delicacy that can be sold on the market for a higher price than standard, raised-for-meat ducks (*rouya*, known in English as table ducklings).

The reason for their higher market value is precisely because of their long life, at least one year of age. One of the most significant consequences of intensive poultry production methods, such as those that accompanied the industrial production of broiler chickens and table ducklings in China, is the shortening of time from birth to slaughter. Indeed, the invention of the broiler or "spring chicken" industry in the 1930s was initially an incidental outlet for excess juvenile males: only a few full-grown roosters had any use. In the U.S. between 1940 and 1995, the average broiler live weight increased from 2.89 pounds to 4.63 pounds, while the time required for a bird to grow from chick to market weight dropped from over seventy days to less than fifty.[63] For table ducklings, those ducks raised exclusively for meat, industrial feeds and breeding reduced growout to slaughter time from seventy-five days in 1928 to thirty-five days in 1993.[64] In China's livestock revolution the same methods of "agro-industrial just-in-time" pioneered in the United States—contract farming, vertical integration of the entire product chain, scientifically controlled breeding—also dramatically reduced the growth times of broilers and meat ducks.[65]

In this context, Chinese consumers turn to Old Mother Ducks—*because of their relatively old age*—for tastes and healthful qualities that more rapidly grown and slaughtered meat ducks are suspected to lack. Often served in soups with a simple ginger broth, Old Mother Ducks are a *bu* (supplemental) cuisine, believed to transmit medicinal or restorative powers. Resembling the growth in artisanal cheese or local pork in the United States, the Old Mother Duck reflects the emergence of consumer food markets in contemporary China based around sentiments of place and quality amid worries about food safety and homogeneity.[66]

Farmers are keenly aware of this added value, and many reminded me that Old Mother Ducks can fetch much more money per weight than common meat ducks. They are well-known delicacies. Yet I was surprised to find that the farmers *themselves* do not see, or taste, these qualities in the Old Mother Ducks that they raise. Although they knew the ducks will become valuable food, they insisted that *to them* these ducks are unpalatable and bad to eat. The poor taste, they explained, is because these commercial ducks are raised on manufactured animal feeds.

The additional farming of ducks "on the side" is, therefore, a renewed form of farming for self-consumption. This farming for self-consumption is nothing like a "subsistence ethic" focused on ensuring the means for a basic living.[67] Instead, even as farmers embark on the business of raising thousands of layer ducks for sale as eggs and meat, they also continue to raise a separate flock of ducks according to a different value: to have tasty duck meat for their own family's consumption on special occasions (e.g., the New Year holiday) and for gift giving.[68] As one farmer explained to me, he raised ducks on the side because only that kind of duck is "good to eat, an authentic local product [*tu chanpin*].[69] They haven't eaten commercial feeds; those that eat commercial feeds aren't delicious."

Similar distinctions have been recently noted by a number of anthropologists studying in rural China. Anna Lora-Wainwright, in an anthropological study of cancer in rural China, describes how farmers willingly use farm chemicals—such as pesticides and fertilizers—on cash crops destined for market sale. Yet they refuse to use these same chemicals, despite their ready availability, for "food intended for home consumption." Although they acknowledge and affirm that these chemicals lead to higher productivity, they never use them on the kitchen gardens and crops they will consume themselves. This difference in practices is associated with a difference in *evaluations*: the farmers say that crops grown in their kitchen gardens, without farm chemicals, taste better. In addition, they attribute cancer to the consumption of cash crops treated with farm chemicals or to working in these fields, suggesting, Lora-Wainwright argues, that they favorably appraise the *healthfulness* of food grown in their own kitchen gardens. Lora-Wainwright also notes that villagers deploy similar distinctions in pig farming.[70]

Gonçalo Santos, discussing a village in Guangdong, remarks on the surprisingly persistent use of human manure, or night soil—and especially urine—as a fertilizer in contemporary China. Before the introduction of chemical fertilizers during the 1960s and 1970s, night soil was used on most fields in China. Today, by contrast, human manure is applied *exclusively* on private kitchen gardens. One farmer cited by Santos explains that "'urine-fed vegetables' taste better than vegetables grown with farm chemicals." This man pointed out that farmers used industrial chemicals on crops grown for "economic" purposes, but most villagers "producing vegetable for self-consumption use 'watered urine' in their small gardens."[71] Ellen Oxfeld also describes a preference for homegrown over market-purchased vegetables among residents of a rural Guangdong village, a preference that she finds is associated with the minimal use of pesticides and chemical fertilizers in these kitchen gardens.[72]

Unpacking the contemporary opposition between "specialized" commercial farming and "sideline" household farming helps show why models of linear restructuring, such as those proposed by FAO biosecurity plans, are unlikely to be fulfilled. So-called backyard farming is not, or is no longer, a consequence of subsistence-based constraints or lack of access to commercial inputs (such as manufactured feeds). Rather, small-scale, by-the-side-of-the-house duck farming persists as a *choice* based on values of producing a certain *quality* of food for the family. These values remain even when the household is engaged in commercial husbandry as well. As a result, small-scale free-grazing or "scattered" (*sanyang*) husbandry is unlikely to disappear, even if access to capital and inputs for restructuring is increased through development plans that extend the reach of "dragon-head" corporations.

Perhaps I could end the discussion there. However, the troubling interactions between the FAO's initial biosecurity plans for restructuring the epicenter and the actual landscapes of duck farming in the Poyang Lake region go further. As I discussed above, duck farmers often house their market ducks in special sheds, frequently at some distance from human habitations. In many respects, therefore, it appears that these specialized commercial farms contain an incipient kernel of biosecure husbandry. After all, enclosure is widely seen as the core of farm biosecurity. Yet I found that the deep insecurity of commercial duck farming in the Poyang Lake region—structured by what I call *vital uncertainty*—limited the consistency of enclosure. Even when farmers raised tens of thousands of ducks in enclosed sheds, that is, *they still brought them to graze in rice fields*. And paradoxically, as I show in the final section, the anticipatory discourses that identify Poyang Lake as a pandemic epicenter only further increased farmers' reliance on free-grazing practices.

VITAL UNCERTAINTY

As ducks are excluded from the household and raised in the specialized husbandry zones, as they come to stand for their market value rather than their edibility, they take on a novel temporal character—they stand as a certain kind of *future* that might come to be: gain or loss, wealth or ruin. Rather than participants in the annual life of the household, the specialized ducks become embodiments of household wealth—merely a moment in the transformation of money into more (or less) money. The identification of market ducks, raised on manufactured feeds, as *nonfood* (according to the farmers' tastes)—and as *stock* rather than *commensal associates*—turns market-oriented ducks into a living embodiment of capital risk.[73]

Of course, all market economy is driven by the uncertainty of investment. However, the making of *living beings* into commodities faces a unique mode of vital uncertainty. A farmer once told me an idiomatic expression, one that, he said, would help me understand how poultry farmers engage with their birds:

家财万贯　　jia cai wan guan
带毛的不算　dai maode bu suan

When counting the household's wealth in strings of cash,
Don't include those with fur and feathers.

This is the traditional view of Chinese farmers, he explained: despite the money that can be made from livestock farming, one can never be certain that animals count as part of the household's wealth. Why? I asked him. Because they are vulnerable to die unexpectedly from disease.

Whether or not the expression accurately represents traditional viewpoints, the uncertainty of placing the household's wealth in the living bodies of domestic animals is a useful key to the practices and affects of duck farmers in the Poyang Lake region today. Alongside the opportunity for gain made possible through the business of duck farming has come the threat of ruin.

FIGURE 3.5. A duck shed in ruins.

The lowlands around the lake are scattered with abandoned duck sheds, literal ruins that exist as visible reminders of farmers who tried out the duck-raising business and failed (see figure 3.5).

As we walked with Chen out to his commercial duck shed, we passed a series of run-down sheds. "They've all gone broke," Chen told me. When farmers spoke about what had changed in the farming of poultry over recent years, they talked about new challenges, about farmers giving up on duck farming, or reducing the size of their flocks, about financial ruin. Almost always they blamed disease. "These days raising ducks is no good," one farmer told me. "Too many diseases. Especially the 'flu' (*liugan*). That liugan, it's hard to cure. For one, it's hard to cure; for another, it's dangerous, it's highly contagious. We really fear that liugan."

Tang, the duck technician we first met in the Introduction on the embankment of the bird refuge, explained to me that "the one thing that has changed: more diseases. Too many ducks, too much pollution." He elaborated: "Generally, if the water is good, then it's fine. [The ducks] won't fall sick. Only if the water source isn't good, if say you raise ducks in a place for five years, or ten years? If the ducks, you raise ten batches, or five batches, the place can't support it; it's too long."

South of Nanchang, Ms. T, a small-scale duck farmer, suggested it was not so much the water quality as the density of ducks that brought diseases. Her duck shed is directly across a courtyard from her house, but facing away from the house toward a large pool of water. She explained that every year, between the ninth and the eleventh month on the agricultural calendar, her flocks have been struck by disease. Because the egg production drops when the ducks are sick, each year she has lost her capital, as feed costs exceeded earnings from the sale of eggs. As a result, over the years she has reduced the number of ducks she raises from eight thousand to three thousand, and this year only raised sixteen hundred. She explained that the diseases arrive because there are too many ducks farmed in this region. "Density of husbandry is too high. Too many ducks."

Agricultural production contains a distinctive mode of uncertainty because human labor is only a partial driver of the production process. The object of agricultural production is a *living being*, so production involves a process of natural growth that is, to a greater or lesser extent, independent of human labor. Marx pointed out that not all of "production time" is made up of "working time"; that is, the time of production is not completely occupied by human labor. In addition to human labor's transformation of objects into products, many spheres of production also contain times "during which the subject of labor is for a longer or shorter time subjected to natural process, must undergo physical, chemical, and physiological changes, during which the labor-process is entirely or partially suspended." Marx suggested that this time of production played an especially important role where production involved processes of fermentation, preservation, or agricultural growth.[74]

The capital or wealth invested in agricultural production depends as much on these nonlabored *growth* processes as it does on the transformative work of human labor. When ducks are designated as carriers of household wealth, this wealth is exposed to a mode of uncertainty that goes beyond the characteristic risks of capital investment. This mode of uncertainty lies at the intersection of biological disease and market institutions. What if the ducks do not grow? What if they will not live?[75]

To be clear, I am not claiming that sideline duck farms are biologically less vulnerable to disease than commercial farms. Many farmers spoke of disease afflicting sideline flocks and of duck or chicken "plagues" sweeping across villages. The novelty of the recent liugan is not that ducks are falling sick for the first time, but rather that in the context of specialized farming sick ducks carry the threat of ruin. It is telling, therefore, that the sideline farming of ducks does not produce the same affect of uncertainty in the face of disease.

When sideline ducks sicken or die from disease, farmers simply slaughter or butcher and consume the birds. Because little or no capital is invested in the lives of the sideline ducks, the household's wealth was never at stake. And although the bird's life might have been shorter than expected, its end was the same: food for the household table. When disease strikes a market flock in the "husbandry zone," on the other hand, the outcome is quite different. In many cases the birds cannot be sold at all, but must be burned, buried, or otherwise disposed of. In some cases, they are exchanged for government compensation but at a severe loss. In other cases, farmers illegally sell them to traders at cut-rate prices.[76]

As farmers aim to mitigate this vital uncertainty, a new domain of practice has emerged. Anthropologists have long noted that although labor has no material impact during the time of production, this by no means indicates an absence of agricultural practices. Algerian farmers frequently visited the fields when there was nothing to do there, Pierre Bourdieu noted.[77] Edmund Leach noted that the "customary procedure" of Kachin rice growing involves much more than "technical acts of a functional kind."[78] Trobriand gardeners had a "surprising care for the aesthetics of gardening," as Bronislaw Malinowski put it.[79] Alfred Gell added that Trobriand gardens "are meticulously laid out in squares . . . according to a symmetrical pattern which has nothing to do with technical efficiency."[80]

Based on these insights, Malinowski proposed a conceptual distinction between "the way of garden work," or technical labor, and "the way of magic," such as the spells incanted over the garden that aimed to assist the tubers in their underground growth. Malinowski's "way of magic," in short, refers to those practices intended to encourage growth without the direct application of labor.[81] We should extend Malinowski's insight beyond what he considered the special domain of the magical. What is at stake is the response to a general problem to vital uncertainty, one that is particularly acute in agricultural production. Growth will always retain a certain heteronomy to direct instrumental rationalization; the production time of raising ducks is longer and different from the time that the farmer spends in direct husbandry work. Disease, which arrives unexpectedly and irregularly, appears as the actualization of the uncertainty inherent when growing living beings for a market. Rather than a distinction between technical work and magic, we can distinguish two domains of technical activity: the first directly encouraging the growth of the ducks and the second aimed at mitigating vital uncertainties.

For example, pharmaceutical treatment of duck diseases is a flourishing business. As I walked from the local bus station on my first visit to the Xiaolan poultry egg wholesale market, I was perplexed to see an enormous, square red banner hung outside the entrance, emblazoned with a single white character: 药 (*yao*, medicine). I thought for a moment that I must have arrived at the wrong place: maybe this was a medicine market? In fact, the egg market is lined both outside and inside with small shops selling medicinal preparations to treat poultry diseases—including both Chinese herbs and "Western" pharmaceuticals in pill bottles. The owners of these shops act as veterinarians, or what I call duck doctors, despite only limited training and frequent absence of official license. As I describe in chapter 6, these duck doctors diagnose disease, sometimes visiting farms or cutting open dead birds, and prescribe treatments.[82]

The boom in livestock pharmaceuticals reflects the growth in vital uncertainty experienced by farmers—that is to say, it indexes the emergence of a new domain of technical practice above and beyond duck husbandry proper. Interestingly, many farmers explicitly marked out the space of pharmaceutical treatment as an area beyond their own experience and practice. Farmers told me that for "small diseases" they had enough knowledge and experience to respond effectively themselves, but for "new" and "big" diseases they relied on the expertise of the duck doctors.

Immunization of poultry is a second new technical practice. In 2005 China's State Council issued a new policy requiring mandatory universal immunization of poultry, including ducks, against the highly pathogenic avian influenza H5N1. With over 5 billion poultry, the enormous ambition of this project is obvious. I discuss the vaccination program in more detail, along with some of the scientific controversies that it engendered, in chapter 5. But how have rural farmers taken up the vaccination program? The necessity of vaccination—although it cannot be said to be exactly universal—is now an everyday and widely accepted part of market duck farming in the Poyang region. Indeed, almost whenever the topic of disease is broached, a farmer will assert that he or she vaccinated the flocks. Chen, for example, explained that all of his market ducks are vaccinated. Tang told me that avian influenza (*qinliugan*) is not a very serious threat because "you can 'hit' [*da qinliugan*], vaccinate [*dazhen*]; it's only if you don't vaccinate that it's not OK."

Along with the material practices involved in the techniques and technology of vaccination, the policy of universal poultry immunization has also transmitted a distinctive mode of uncertainty from the state to the rural farm: an orientation toward the future based around the idea of *prevention*. The slogan "Make Prevention the Priority" is painted on walls and printed on posters across the countryside. This slogan was popular in Mao-era public health, such as in the antimalaria and anti-schistosomiasis campaigns of the 1950s. These

FIGURE 3.6. A vial containing vaccine for HPAI H5N1.

campaigns, initially conducted during a time in which biomedical resources were scarce, relied on mass mobilization to transform environments and remove dangerous pests. In the Poyang Lake region, for example, collective labor constructed dykes, drained marshes, and removed the snails that were implicated in the transmission of schistosomiasis.[83] By the 1960s and 1970s, a new configuration that linked pharmaceutical innovation, industrial manufacture, and free drug distribution transformed the meaning of prevention and led to the successful control of both diseases.[84]

In the case of avian influenza, prevention has been concretized in the vaccine vial and needle, producing a novel integration between biomedical expertise and mass mobilization. The H5N1 vaccines are developed at the Key Laboratory for Avian Influenza in Harbin, one of the most sophisticated animal disease research labs in China and a central actor in the state-led response to avian flu. The vaccines are manufactured at a series of rapidly growing biotech companies. These vaccines are then distributed to farmers through county-level veterinary stations. As a result, farmers in the Poyang Lake region speak of vaccination as "doing prevention" (*da yufang* or *gao yufang*) (see figure 3.6).[85]

When I most recently returned to the Poyang region, I found that a third technology for managing vital uncertainty has now emerged. A farmer told me that he had signed up for "duck raising insurance" (*yangya baoxian*). Over the past couple of years, the government had permitted insurance companies to sell policies to duck farmers because of the threat of disease. A payment of one yuan per bird, for example, would merit a payout of three yuan if the bird died from disease. Duck disease, like the ducks themselves, now circulated on a market.

PROBLEM AMPLIFICATION

Has pandemic-preparedness planning been translated into trajectories of biosecurity development in places like Poyang Lake? On the one hand, a combination of disease outbreaks and biosecurity interventions such as forced culling or market closure did drive a large number of smallholder farmers out of business.[86] At Poyang Lake the combination of market and disease pressures left behind the ruined duck sheds that I described above. This collapse of smallholder poultry enterprises intersected with China's policies for scaling up production. According to China's 1996 national poultry census, more than 104 million households raised poultry, and 96.8 percent raised 50 birds or less. By 2005, after the first wave of avian influenza outbreaks, there were only 34.6 million households raising poultry, and nearly half of all poultry were raised by "scale" farms with more than 1,000 birds.[87] By 2009, only 30 percent of poultry farmers raised fewer than 2,000 birds, and more than 200 farms raised more than 1 million birds per year.[88] However, in this chapter I have suggested that a closer look at this trajectory in the Poyang Lake region reveals a much less linear process of transition. Indeed, abandoning production is not the only way that farmers respond to new uncertainties and pressures.

Consider the way that farmers talk about avian influenza. Many farmers carefully distinguish between *liugan* ("flu") and *qinliugan* ("avian influenza"). Ms. T, for example, complained that her ducks had been afflicted by liugan toward the end of every year, but she denied that she had ever had a case of qinliugan.[89] At first this was perplexing to me: ducks are an avian (i.e., *qin*) species, so wouldn't any duck liugan be, by definition, qinliugan?

One farmer finally enlightened me. In addition to disease and market pressures, he said, farmers struggled with the impact of what he called "the greater environment." Qinliugan, he explained, exemplified such recontextualizing forces. Beyond referring to actual outbreaks of a specific disease, the term carried additional senses: it also embodied the aura of popular fear and state biosecurity interventions associated with the prospects of pandemic flu. When qinliugan broke out anywhere in China, he continued, the actual infection of poultry flocks with the virus was by no means the most devastating impact on farmers. Rather, farmers suffered because of government interventions and consumer reactions to the publicity of the outbreak.[90] The government legally restricted trade in poultry, closed poultry markets, and enforced the massive slaughter of birds. Compensation was absent or inadequate. Consumers, for their part, suddenly refused to purchase or eat any poultry.[91] Many farmers in the Poyang Lake region had been ruined when the government closed down live-poultry markets and sealed off all poultry trade in response to an outbreak of avian influenza, as a trader at the Nanchang live-poultry market also told me.

But unexpectedly, this intensification of vital uncertainty appears to encourage, rather than limit, the practice of free-grazing ducks. Recall that it is the remainder of production time, its excess over directed husbandry work, which opens up the distinctive vital uncertainty of commercial duck farming. Further, the scale of that uncertainty is shaped by the interplay of two forces: market prices (for the cost of feed and the earnings from eggs or ducks) and disease (causing lower productivity or even death of the birds). As a result, it is possible to reduce uncertainty not only by managing the threat of disease (through vaccination or drugs, for instance) but also by managing the scale of market uncertainty. Duck-raising insurance is a great example: it does nothing to treat the diseases that ducks suffer, but it does manage the *uncertainty* of duck disease by transforming it into a financially insurable risk.

Free-grazing is a practice that manages uncertainty by reducing market exposure rather than directly protecting the health and life of the ducks. Duck free-grazing allows farmers to reduce costs spent on feed, reducing the amount of household wealth that is staked on the life of the ducks. Therefore, as duck farming has increased in intensity and scale, and although market ducks tend to be held in the semi-enclosed spaces of the husbandry zone, free-grazing practices did not disappear. The only thing that changed was the form and tempo of free-grazing. Ducks raised in a "specialized" commercial mode, unlike sideline ducks, are no longer left to graze throughout the day and throughout the year. On the contrary, market ducks are kept in specially built sheds and fenced enclosures for most of the year. Then, during intensive grazing phases—usually immediately after harvest or before planting—the farmer drives ducks to rice paddies throughout the region to glean unharvested rice and insects (see figure 3.7).[92] Farmers rely on these intensive grazing phases to reduce the overhead costs of manufactured feed and ensure that the sale of "spent ducks" will be profitable.[93] Unexpectedly, the intensification of vital uncertainty caused by the global response to the threat of pandemic influenza

FIGURE 3.7. Grazing ducks in a rice paddy after the harvest.

may drive poultry farming practices along a trajectory that is at odds with plans for biosecurity development.

The identification of rural China as an epicenter of pandemic influenza is grounded in the apparent danger of free-grazing ducks, whose promiscuous movements and contacts with other species could lead to the emergence of new flu viruses. As a result, when new avian viruses such as H5N1 emerged, global health agencies declared free-grazing ducks to be a key risk factor and proposed biosecurity interventions, such as culling, market closures, and improvements in biosecurity, to manage risk. When these biosecurity interventions increased the uncertainty of duck farming in the Poyang Lake region, however, farmers saw free-grazing as a crucial mechanism for reducing what I have called their vital uncertainties and protecting their investments. A positive feedback loop began to take shape, a circuit in which biosecurity interventions and farming strategies amplified each other and new uncertainties appeared.

By the end of 2011, as I finished my fieldwork in China, the FAO was becoming aware of these displacements to its initial biosecurity models. An FAO report on "lessons learned" from the HPAI outbreak featured a discussion of biosecurity planning and its unintended outcomes. "In general," the authors of the report acknowledged, "considering the complexity of, and variation within, the commercial poultry industry throughout the region, solutions developed locally through direct engagement with stakeholders have been more effective than solutions imported from other regions."[94] In the next chapter I consider the additional displacements that took place when scientists did directly engage with the inhabitants of the pandemic epicenter.

CHAPTER FOUR

WILD GOOSE CHASE

On a December day in 2007, an American wildlife veterinarian named Scott Newman stood in a rice paddy near the shore of the Poyang Lake. He looked up as a flock of swan geese lurched into flight and circled in the sky overhead (see figure 4.1). *Anser cygnoides*, he thought to himself. Swan geese. Newman knew these birds. Both a doctor of veterinary medicine and a PhD in wildlife and disease ecology, he had worked for the Wildlife Health Alliance, the Wildlife Trust, and the Wildlife Conservation Society. Hired by the FAO in 2006 to coordinate international research on the role of wild birds in avian influenza, he had brought a team to Poyang Lake to study how long-distance bird migration might drive the emergence of influenza pandemics.

Newman and his team planned to surgically attach transponder tags and antennas to migratory birds, including swan geese, in order to track them by satellite both at Poyang Lake and when they migrated north in the spring. The team designed the study to better understand how, where, and when wild birds come into contact with domestic poultry. According to their hypothesis, contacts across what they called the "wild bird–domestic poultry disease interface" may encourage the reassortment and emergence of pandemic flu viruses. Drawing on lengthy traditions in wildlife management and wildlife health, the study focused on a boundary that separates two forms of animal life: the wild and the domestic.[1]

FIGURE 4.1. Farmed swan geese (*dayan*) in flight.

As he watched the flock of swan geese flying above him, Newman felt puzzled, he later recounted to me. Swan geese, he knew, are an endangered species of wild waterfowl. But according to what a poultry farmer had just told him, the geese above him were not wild. Indeed, these swan geese belonged to the farmer, who had bred, incubated, hatched, raised, housed, and fed them in preparation for slaughter and sale. Then again, Newman wondered whether the birds could really be considered domestic. They were hardly tame. And the farmer kept insisting on their *wildness*, their delicious wild taste. To validate their authentic wildness, the farmer prodded Newman to watch them in flight, reminding him—of course he hadn't forgotten—that domestic geese can't fly. Newman agreed that they resembled wild swan geese so closely that even true wild birds would be unable to recognize the difference.

Newman had come to Poyang Lake to study the interface between wild and domestic birds. But he found that poultry breeders were actively recomposing the material qualities of wildness and domesticity in their husbandry of swan geese. Despite the apparent contradiction in terms, they were breeding wildness.

This chapter explores the displacements to scientific practices that occur when scientists move experimental systems into the working landscapes of the pandemic epicenter. Despite the universal scope of validity claimed by scientific facts, historical and ethnographic studies of scientific practice demonstrate that the content of knowledge is shaped by the local spaces of inquiry.[2] At the same time, the vast majority of historical and ethnographic studies of the sciences have focused on one particular kind of scientific venue: the laboratory. No doubt, the historical emergence of the laboratory is central to the rise of modern science and medicine.[3] Equally important, the laboratory has been a strategic site for studying science in the making because fieldwork inside labs makes visible the concrete devices and equipment that undergird the construction of global facts.

Based on specific empirical studies of scientific practice, laboratory ethnographies developed a model of scientific knowledge production that Karin Knorr-Cetina has aptly characterized as *detachment*: "Laboratories rarely work with objects as they occur in nature": the epistemological advantage of laboratory work lies in "the detachment of the objects from a natural environment and their installation in a new phenomenal field defined by social agents."[4] *Detachment* is an incisive term because it captures the spatial movement of material objects from their natural environments into the lab and it also describes how the laboratory reinforces a sociological separation between scientific experts (inside) and other lay actors (outside). Scientists work to maintain their detachment from laypeople—both defending their distinct professional status and protecting themselves from unwanted influences or bias. The laboratory is the tool that enables the interrelated detachment of both objects and subjects of science. Understanding the lab as a tool rather than a place, anthropological and sociological accounts of science extended the idea of *laboratory practice* beyond the "houses where experiments take place."[5] The procedures and protocols of detachment that characterize laboratory practice—for instance, the collection of inscriptions or standard traces—came to stand for scientific practice more generally. "For the world to become knowable," Bruno Latour has argued, "it must become a laboratory."[6]

As some historians have emphasized, however, the laboratory has been only one among many sites of scientific inquiry, and differences of place do matter for the ways in which knowledge is produced.[7] One area that has become a growing focus of interest is the distinctive epistemological and cultural characteristics of the field sciences. Modern science in the field takes shape in the shadow of the historical ascendancy of the laboratory as the dominant form of scientific practice. As the laboratory rose in epistemic status, therefore, "'the field' was simultaneously reconstructed as the residuum of messy, complex, and uncontrollable nature."[8] Science in the field is distinctive, Henrika Kuklick and

Robert Kohler argue, because "unlike laboratories, natural sites can never be exclusively scientific domains."[9] Laboratory ethnographies exposed the practical work conducted by scientists to exclude society or environment from the controlled space of the lab. Science in the field, by contrast, necessarily takes place amid what environmental historians have called a "working landscape," a setting already being produced and reproduced for a wide variety of human ends.[10] Because the objects of field sciences are studied in situ, the work of building authoritative knowledge must respond in one way or another to the other inhabitants of the field site.

Because of this residual history of field research, there is a tendency to distinguish and classify the sciences into lab and field varieties. Yet as Robert Kohler shows, the boundary between laboratory and field, though often "well patrolled," is highly permeable: "The cultural geography of field and lab is like a biogeographic ecotone, an area where two biota intermingle and where neither has a clear advantage, and where we expect to find the odd hybrid."[11] This chapter explores the changing trajectory of influenza research as it moves from labs into the fields of the hypothetical pandemic epicenter. It follows Kohler's call to attend to "border crossings" between lab and field spaces, exploring how new knowledge is made as experimental systems move out from laboratory centers. Extending the model of laboratory detachment to explain all forms of scientific practice has resulted in a monomorphism that neglects other forms of scientific work and development. Here I describe a trajectory that I call field displacement, in which poultry farmers, the producers of this working environment, reshape the form of a scientific research object in unexpected ways, creating in the process an "unprecedented event" that changes the course of scientific knowledge.[12]

VICTIMS OR VECTORS

A 2005 outbreak of avian influenza A H5N1 on China's northwestern plateau crystallized the movement from laboratory to field around the figure of the migratory bird. During that spring, Chinese park rangers found thousands of dead birds on an island in the middle of the remote Qinghai Lake. According to Scott Newman's appraisal, it was "the single largest H5N1 wild bird mortality event that has ever occurred."[13] In its sheer scale the Qinghai epizootic indicated sustained transmission of the virus among wild birds. Later that same year, several dead birds, infected with the H5N1 strain, dropped out of the sky over Europe. Pandemic flu researchers began to wonder if wild birds might play an unexpected role in the long-distance transmission of highly pathogenic flu viruses. As Newman and his collaborators aptly phrased it, everyone wanted to know whether wild waterfowl were "victims or vectors" of the virus.[14]

As funding for wild bird studies grew, the FAO hired Newman to coordinate international research on the role of wild birds in avian influenza. In 2006 he helped organize a study of wild bird migration at Qinghai Lake, along with collaborators from the U.S. Geological Survey and the Chinese Academy of Science. In these studies they attached satellite transponders to birds at Qinghai Lake and observed their migratory flight paths across the Central Asian Flyway. The team hoped to understand how flu viruses got into the remote Qinghai Lake and where they might go next. For example, they tracked ruddy shelducks and bar-headed geese as the waterfowl migrated over South Asia. In another study they tracked whooper swans to Mongolia, where a second large migratory-bird die-off occurred in 2006. Through detailed analyses of "temporal and spatial relationships" between the swans and domestic poultry, the team concluded that whooper swans were sentinel species but probably not vectors of influenza viruses.

At around this time, several papers published by Chinese and Hong Kong virology labs brought attention to Poyang Lake, far to the east in the intensively farmed plains of the Yangtze Delta, and located within the parallel but distinct East Asian–Australasian Flyway. In one paper, Guan Yi and colleagues from Hong Kong and Shantou reported the isolation of avian influenza A H5N1 virus from several living, apparently healthy wild birds during routine sentinel surveillance at the lake. Because infected wild birds were expected to sicken and die quickly when infected with highly pathogenic forms of flu, this finding confirmed the hypothesis that migratory birds played a role in the long-distance transmission of highly pathogenic flu viruses. The study also did something else, however: for the first time, it identified Poyang Lake as a hypothetical pandemic epicenter. As the authors put it, "Migratory ducks at Poyang Lake could have survived infection with the HPAI H5N1 virus and transmitted the virus over long distances during migration. This possibility may provide insight into reported H5N1 outbreaks in Mongolia, Siberian Russia, and Europe that have been linked to migratory birds."[15] A second paper, a computational work in viral genomics, claimed that the origins of the first H5N1 virus detected in China (known as Goose/Guangdong/96) could be traced to predecessor viruses isolated in the Poyang Lake region. After reading these papers, Newman and his team decided to move several research initiatives to Poyang Lake. As Lei Fumin, an ornithologist from the Chinese Academy of Sciences, explained the rationale behind the move, "Qinghai strains can be traced to one early strain from Poyang based on the genomic analysis."[16]

Newman visited Poyang Lake while attending the 2006 International Living Lakes Conference, an environmental conservation meeting held in the provincial capital, Nanchang, just south of the lake. Speaking on a panel devoted to "Avian Influenza, Wildlife, and the Environment," Newman discussed "integrated fish farming" in China as a possible mechanism for virus transmission from domestic poultry to wild birds. These integrated fish farms "directly use fresh poultry waste as a production input"—that is, as an inexpensive fish feed. High quantities of virus are known to be excreted by birds infected with the H5N1 strain of influenza. Newman concluded that "in such systems with little or no biosecurity measures in place, the likelihood of multiple wild species interaction and possibility of disease . . . transmission could be considerable." The identification of a landscape feature or system—the fish farm—as site of "species interaction," and therefore a *pathogenetic zone*, is worth noting here. In chapter 2 I showed how Kennedy Shortridge's strictly virological sampling of the pandemic epicenter gave way to what Marius Gilbert and others called "eco-virological" approaches to mapping pathogenetic landscapes. Newman's work at Poyang Lake was in many ways intended as a "model" of eco-virological research, or what he sometimes called a "One Health approach."[17] Rather than analyzing and classifying viruses in the laboratory, Newman planned to investigate the ecological relationships that contributed to the transmission of flu viruses into new populations.

At Poyang Lake he joined a team of researchers funded by a grant from the U.S. National Institutes of Health (NIH) that was developing an "integrated pilot study" that drew together a wide range of disciplinary perspectives. The pilot study included spatial analysis of landscape and land use derived from satellite imagery, surveys of domestic poultry farming density and market chains, tracking of wild bird migration, and sampling of wild and domestic birds for viruses. Population ecologists, livestock veterinarians, geo-spatial analysts, and socioeconomists joined Newman and other wild bird specialists. Participants referred to this informal research collective, in shorthand, as the "NIH group," indicating the provisional, temporary (grant-specific), and noninstitutional character of the collaboration.

Members of the NIH group, like Newman, stressed the importance of an "ecological research perspective."[18] In publications or conversations these researchers lamented that expertise in the ecological sciences had too frequently been missing from existing virology studies of avian influenza, ironically so when these studies made reference to the importance of ecosystem connections. In a programmatic paper the NIH group criticized previous studies conducted at Poyang Lake for a "lack of detail in identifying migratory waterfowl to species level [that] precludes analysis of ecological aspects of the disease."[19] At issue were several H5N1 viruses isolated from wild birds during viral surveillance programs at Poyang Lake. As John Takekawa, the lead author on the paper and a wild bird migration specialist, explained,

> A lot of the [virus] sampling has been done without designation. Now, it's fine if you can get to species level, it's better than you started; I mean initially it was just like "duck." And there's like huge differences in species, right? And so all this is to us [wild bird specialists] common in that you look at a bird and you know that "Well, that's a different bird, and it's different from this one over here, 'cause it's doing this bit of behavior, it's completely, it's not going to be found in that habitat, all of those things you automatically know, and you hardly think about it, you don't realize that over there a virologist is thinking, 'That's a duck.'" You know? A tree's a tree. And that redwood and that oak tree, it's all the same.

To correct this absence, the NIH group designed a pilot study to "integrate" a diverse array of studies underway at Poyang Lake around a common research object: what they called the "wild bird–domestic poultry disease interface." They proposed that the transmission of viruses across the wild-domestic interface was a "key factor integral to the evolution of LPAI [low pathogenic avian influenza] into HPAI [high pathogenic avian influenza]."[20] As viruses trafficked across the interface, according to this hypothesis, they changed from low to high pathogenicity, gaining pandemic potential. By describing the contours and pathways of the wild bird–domestic poultry interface at Poyang Lake, the NIH group suggested they might be able to trace the route along which avian flu viruses emerge into pandemics.

At the same time, Newman and his colleagues relied on mechanisms of detachment—what I have called laboratory practice—to achieve this integration of their widely diverse modes of research. Rather than isolating viruses and bringing them back to the lab for analysis, the NIH group collected a different kind of "trace": spatial and temporal coordinates. Each kind of observation conducted at Poyang Lake—whether remote sensing of landscape and land use, tracking of wild bird migration routes, or surveys of domestic poultry populations and movements—was tagged with geospatial and (usually) temporal markers. In doing so, every observation could be located within a single, standard, spatial and temporal grid.

I first began to recognize the importance of these spatial marking practices in an incidental conversation with Professor Wang, a geographer involved in the international flu research program that I was following at Poyang Lake. I had just returned to Nanchang after visiting duck farmers in a rural part of the region. Wang, who had helped connect me to the farmers, was waiting in a café adjoining my hotel. I began showing him some pictures and telling him about what I'd learned from the farmers, when he suddenly interrupted me: "You captured the GPS coordinates for these photos, didn't you?" he asked. I replied that I hadn't. "Ah, too bad," he smiled, almost forlorn. He explained that all of the diverse forms of data that he and the NIH group collect, from poultry surveys to virus samples, from wild bird tracks to seemingly irrelevant photographs of wetland landscapes—are stamped with a geographical coordinate and usually temporal marker. Indeed, he noted that they use special cameras outfitted with GPS devices for precisely this purpose, and urged me to purchase one for myself.

I then recalled a story that our van driver, Mr. Xiong, had told me a few days earlier. On one trip, Mr. Xiong took graduate student Zhao all across the Poyang Lake region, following any roads they could find—including those not on the map—in order to look for poultry farms. Whenever they saw a poultry farm, they stopped and conducted brief interviews with farmers about husbandry practices (How many birds? When and where do you free graze them?). If no one was there, Zhao would make an eyeball estimate of the flock. Most important of all, Mr. Xiong explained, was a portable satellite GPS tracker that sat between the driver and the passenger seat of the van. At each stop, alongside the brief handwritten notes that he entered into a notebook, Zhao carefully recorded the GPS coordinates.

Without these tags of spatial and temporal position, Professor Wang explained, there would be no way to integrate different forms of information into a unified model of the ecology of influenza. But *with* the coordinates tethered to the other forms of data, the ecological models reveal the domain that overlaps, in space and time, between the wild and the domestic. As the NIH group explained in one study publication, "The main objective of this [interdisciplinary] synthesis was to characterize the temporal relationships between wild and domestic ducks. . . . We compared the temporal pattern of free-grazing domestic duck farming in the Poyang Lake area with the seasonal presence of wild ducks in paddy fields to identify a period with potential increased risk of AIV transmission."[21] The publication includes a "conceptual model of integrated study components," a flowchart showing how data from remote sensing, a domestic duck farm survey, and bird telemetry are integrated to produce an original object of analysis, highlighted in a gray box with the label: "AIV Transmission between Domestic and Wild Ducks in South China Agroecosystems."[22] In this study, pathogenesis is located not in the virus but in the "pattern," "presence," and "period" of multispecies interface—a space and time of transmission. The concept of the wild-domestic interface, therefore, relied on and reaffirmed an interdisciplinary model of work, one in which each discipline contributes its own specialist data to a common model of overlapping movements and presences. Interdisciplinary synthesis aimed to capture and represent multispecies interface. But it also relied on the shared mechanism of detachment—spatial and temporal coordinates—that enabled these diverse forms of data to inscribe a common object.

In the original formulation of the study, the "interface" was understood as a spatial pattern and temporal period in which two distinct species populations—wild migratory birds and domestic waterfowl—came into contact. But when Newman arrived to conduct his studies at Poyang Lake, he found himself watching a group of birds that couldn't quite be placed on either side of the interface: Were these wild geese, because they could fly and were phenotypically indistinguishable from wild swan geese? Or were they domestic geese, because they were bred and raised on a farm?

The NIH group had derided virologists for their inability to recognize the differences between mallard and pintail ducks. They came to realize that their own distinction between wild and domestic kinds of bird was equally inadequate.

BREEDING WILDNESS

Just inside the large embankments that keep the floodwaters of Poyang Lake at bay, Wang Fenglian raises his wild swan geese (*dayan*). I first visited his farm in the summer of 2011, brought there by one of the NIH group's Jiangxi Province collaborators, and visited or stayed with them frequently afterward. The farm is located on land belonging to Shibahu ("Eighteen Households"), a small village nearby that was built on land reclaimed from the lake by earthwork dykes and embankments in the 1950s. Wang and his family formerly lived in a much older village about five kilometers down the road. Born into a family of rice farmers, Wang had risked a wide range of enterprises since the beginning of economic reforms: he had raised fish in a small pond; he had bred dogs; he had even gone into business in the provincial capital. Some of these enterprises had brought profits, others great losses. In 1999, Wang told me, a friend showed him an article about somebody in Jiangsu or Zhejiang who was farming wild swan geese, and Wang sensed an opportunity. Wang recalled the annual arrival of migratory swan geese in his childhood and knew that they were highly valued as food. Yet this supply relied on hunting, which, as environmental laws

FIGURE 4.2. Wild swan goose farm.

strengthened, had now been banned. In 2001 he contracted land from Shibahu to begin breeding wild swan geese and incorporated the Po Lake Wild Animal Breed Company Ltd. The moderately sized plot of land contains a house where Wang, his family, and employees live; a few sheds for the wild geese and ducks; a special building containing an industrial egg incubator; and a pond where the birds often swim. Today, Po Lake ranges among the largest wild bird farms in the area (see figure 4.2).

China's post-Mao reform policies simultaneously expanded wildlife conservation and promoted agricultural commercialization, trends that frequently came into direct conflict.[23] At Poyang Lake a large section of wetland was set aside as a migratory-bird refuge in 1983, and other sections were designated as an "agricultural production base" focused on duck breeding. The double goal was captured by a slogan I saw written on the wall of a building in the migratory-bird refuge: "Protect birdkind, enrich humankind!" (*baohu niaolei, zaofu renlei*) (see figure 4.3). As early as the 1980s, Jiangxi Province officials suggested that wild bird breeding could help resolve conflicts between social and ecological interests by meeting demand without poaching from the wild.[24] The breeding of wild swan geese began to grow rapidly about a decade later,

FIGURE 4.3. "Protect birdkind, enrich humankind."

encouraged by expanding elite consumption. An exemplary article, published in a Henan Province agricultural extension journal in 1999, promotes the activity as a timely response to unprecedented markets in the quickly growing coastal cities: "Swan goose is a special poultry that our nation has only recently begun to breed from the wild [*xunyang*], and in some coastal cities there is a rather large market for its consumption. . . . As a result, the prospects are good for the development of swan goose breeding."[25]

A Chinese newspaper has described the rapid increase in the breeding of wild animals in the past two decades as a contemporary "fever," drawing on a trope often used to depict cultural trends of the post-Mao period.[26] The consumption of wild flavor, or wild-animal food products, grew rapidly across China during the 1990s and early 2000s.[27] The pursuit of "rarity" and wild flavor in food consumption is closely linked with practices of symbolic distinction in business, government, and social banqueting. After the SARS outbreak was linked to the farming and marketing of live civet cats, China's government implemented several new regulations and restrictions on wild-animal farming. However, "wild consumption" remained an important part of China's banquet cultures, as well as a crucial strategy for some smallholder farmers.[28]

As I explained in the previous chapter, the introduction of market mechanisms into the livestock sector during the early 1980s set in motion a transformation in the social and technical modes of animal production. When "households" consolidated as economic units in the aftermath of decollectivization, livestock farming—and in particular poultry husbandry—quickly became an important source of rural livelihood. The sideline farming of poultry expanded and intensified as farm households became "specialized." During the 1990s, however, large industrial poultry enterprises organized as vertically integrated "dragon-head corporations" began to increase market share. Dragon heads are particularly dominant in broiler chicken and table duckling production because of the efficiency of feed conversion made possible by special imported hybrid poultry breeds. Following the growth of the dragon heads and the emergence of new diseases such as HPAI, statistics show a rapid drop in smallholder poultry farms.[29] Many farmers either joined industrial enterprises as contract farmers or left poultry farming altogether. However, during my fieldwork I discovered a third possibility that could be described as a minor contemporary transformation in livestock modes of production.[30] Many smallholders did not abandon poultry production altogether but instead now specialize in local or unusual breeds. One manner of specializing production, which aims to meet the growing demand for distinctive foods among wealthy elites, is to breed and husband wild animals.

China's administrative system categorizes wild animal farming as "special type husbandry" (*tezhong yangzhi*). This category defines wild animals bred under human management as still wild, placing them under the jurisdiction of the State Forestry Administration rather than the Ministry of Agriculture. As an article in the newspaper *Peasant Daily* explains, "Wild animals, even when they are under conditions of human-directed husbandry, no matter how may generations they have been raised, as long as they have not passed through human-directed cultivation [*dingxing peiyu*], nor produced new hereditary characteristics, that raised animal still is classified as a wild animal, and cannot be called a domestic poultry or livestock."[31]

Wild goose breeders take a reflexive approach to the distinction of wild and domestic. Indeed, wild goose farms are engaged in a kind of human-directed cultivation, but one that aims to reproduce the wildness of their birds rather than to tame them. For them, wildness (*yexing*) cannot be presumed as a stable characteristic that would passively maintain itself. Instead, they made the traits of wildness into direct objects of hereditary cultivation.

At Wang's farm I saw the effort the breeders put into demonstrating the wildness to their clients, many of them prospective wild goose farmers themselves.

When buyers express interest in purchasing chicks, Wang invites them for an inspection of the farm, describing the visit as a complimentary course of instruction in husbandry techniques. Upon arriving at the farm, the prospective buyers first eat a sumptuous lunch prepared by Wang's wife, often including a soup made from wild goose meat. Wang's wife told me that she boils the soup with only salt and ginger to preserve the natural "grass flavor" in the meat, a result of the geese foraging in the marsh reeds. She noted that when she cooks chicken, by contrast, she has to add all kinds of *weijing* (a flavor enhancer, such as MSG) to make it even palatable.

The prospective buyers then tour the goose sheds and ponds, where Wang shows them how feeds are prepared. Most importantly, Wang points out the wild qualities in the birds and makes sure that the prospective buyers see the birds in flight. Wang's son, Haohua, acknowledged that if the prospective buyers did not visit the farm, they might not believe the birds were actually wild. Lack of social or impersonal trust is widespread across China, particularly in the food sector, and many consumers fall back on personal trust to ensure food quality.[32] As Haohua quipped, quoting a popular saying about frauds, "Hang up a sheep's head outside the shop, but sell dog meat" (*gua yangtou, mai gourou*).

Wildness, according to the Wang family, is embodied primarily in three traits: general external appearance, such as coloring and shape; the absence of a growth on the base of the beak that appears on domesticated geese; and, above all, the ability to fly. Promotional materials, including the Wang family's website, pamphlets, and packaging materials, draw a close symbolic connection between the birds' ability to fly and their wildness. For example, one pamphlet praises "wild taste," while images of bred wild geese in flight are cut and pasted over pictures of undeveloped sections of Poyang Lake.

In addition to this symbolic work of marking wildness, though, the Wang family is also careful to ensure that their geese physically embody the traits they identify as wild. And this is not as simple as selecting a species of goose from the wild or one broadly categorized as wild and then raising it on the farm. Wang found, to his chagrin, that after four or five generations of human breeding, the geese lose their distinctive wildness, growing knobs on the base of their beaks and losing their ability to fly. His son described this loss of wildness as degeneration or regression (*tuihua*). As a result, techniques of cultivating wildness lie at the center of the Wang family's breeding practice.

First, although the geese are allowed to breed on their own (*ziran peiyu*), the Wangs use netted confinements to carefully manage which birds are breeding. They blame the degeneration of the geese in part on inbreeding, that is, the reproduction of offspring in sexual relations between individuals too closely

FIGURE 4.4. Wild swan goslings, recently hatched from incubated eggs.

related. In explaining their practice to me, Haohua drew on an analogy of the incest prohibitions in classical China: those of the same family line cannot have sexual relations if they are within three generations of relatedness. The geese are divided into families, and during breeding seasons the male offspring are kept in pens separate from their ancestral family. Furthermore, a distinction is made between "pure-breed wild" (*chunzhong yesheng*) birds, which are used for breeding for up to three generations, and the "commodity" birds that are sold as meat. Any birds with noticeable imperfections or problems are also removed from breeding stock (see figure 4.4).

Occasionally, the Wang family also acquires a small number of wild swan geese from the Poyang Lake Migratory Bird Reserve. Using a special license, and drawing on connections at the reserve, the Wang family takes a few geese captured or held there. These birds contain what they describe as "outside" genes (*wailaide jiyin*) that, introduced into the breeding stock, help protect the wildness of the birds.

At the same time, the Wang family also works to enhance the wildness of the geese by managing the influence of the environment. In a promotional brochure that I helped hand out at the China International Forestry Exposition in 2011, Wang describes such environmental management as his innovation. "Our company courageously seeks innovation [*chaungxin*]," the brochure reads, "bravely explores frontiers, in the whole nation the first to free-graze wild geese and wild ducks in the natural wild." In addition to the main farm, the Wang family also established what they call an experimental base much closer to the shore of Poyang Lake's open waters. In a low wetland near the lake, they built four low sheds to house the geese. In front of each shed is a long pond where the geese can swim and graze for insects. A village collective built these ponds for raising fish, and Wang Fenglian reserved rights to them after fish farming declined. Whereas the main farm is on the inside of the embankments that keep flood waters from human settlement, the experimental base is on the outside of the embankment, exposed to the lake's untempered force. One year, rising waters even flooded the experimental base, causing some geese to escape and never return. As Haohua put it, by compelling the birds to accustom themselves to a wild living environment, their wildness will grow and intensify. Or to quote again from a promotional brochure, "To make the human bred swan geese . . . freely move in the waters and wetlands of the Poyang Lake, to allow them to graze for wild foods, will preserve their natural wildness [*tianran yexing*], their ability to fly, and their external appearance."

For the Wang family, wildness was not defined as that which was outside of human touch. Neither was it an internal characteristic of certain individual birds or bird species, as if wildness persisted indefinitely in the organism or the species, no matter the context (as tigers born and raised in the zoo are sometimes still considered to be "wild animals").[33] For the Wang family, wildness was a collection of qualities that could be cultivated or lost. The Jiangxi Po Lake Wild Animal Breed Company was in the seemingly paradoxical business of farming wildness. To them, wildness could even be described as an innovation.

CREATIVE ECOLOGIES

On his business cards, Wang Fenglian refers to himself as chairman of the board of his wild bird breeding company. Although the simple farmhouse where he based his operations may seem discordant with the terminology of the corporate firm, these are not private delusions. Wang has been featured in newspapers and television news programs, as well as in journals such as *Qiyejia* (*Entrepreneur*) and *Xin Shengyi* (*New Business*). These journalistic accounts articulate the character of a man who "had been a peasant" before beginning to experiment with various forms of commercial animal husbandry, frequently referring to his struggles to innovate. Ann Anagnost has analyzed the developmental narratives of the post-reform period that situate the peasant as a passive object to be turned into an active subject of the nation.[34] Wang Fenglian, according to these published accounts, is advancing on the appropriate developmental track, although to be precise the metric of advance is not wealth or civilization, but creativity.[35]

I spent most of my time at the farm accompanying Haohua, who organized the day-to-day management of the farm. Haohua had graduated from Jiangxi Normal University with a degree in athletics a few years earlier. Upon graduation he attempted to become a prison warden, following a good friend's example, but did not pass the examination. He then took a job as a fourth-grade math teacher but left after only one year, frustrated by the obsequious social world of the school leaders and local officials. He returned home to work with his father hoping, he explained, to construct a life "between city and country" as a rural "creative innovator" (*chuangyezhe*). I was surprised by the way Haohua described his farmwork in language that seemed drawn from the world of Silicon Valley and the new spirit of capitalism.[36] For him, farming wild birds was about branding and marketing, not just tending to animals. Farming wild birds was a vocation chosen in order to be in control of one's time, in order to create new things, new products. Haohua often contrasted this life of innovation with state-based employment, and in particular the education sector. This was not only because of his own brief stint as a teacher but also because his girlfriend was an instructor at an elite kindergarten in Nanchang. She was constantly threatening to break up with Haohua because she viewed his work on the farm as unstable and associated with the "low quality" of peasant life.[37] But Haohua countered her by figuring their different careers in other terms. For him, his work relied on risk taking and creativity, in contrast to the stifling security of government jobs. Indeed, he joked, the walls of his girlfriend's security were literal and physical: she could rarely go out with him, as she lived with her colleagues in a work-unit–provided building with curfews and rules about visitors.

The ethos of innovation that Haohua and his father both cultivated, each in his own way, shaped the material practices and economic strategies on the farms of the Po Lake company. Indeed, they kept making new things. When I returned to Jiangxi in late October 2011 after some time at the Emergency Center in Beijing, I found that an entirely new venture now attracted their attention. Haohua picked me up in Nanchang in his brand-new Honda Accord and told me he was on his way to the Quality Inspection Bureau, also in the provincial capital. As we set off, Haohua called his father to get directions to the bureau. We took the roads leading out of Nanchang center, while Hao-hua shouted into the phone, occasionally taking erratic turns. "Be-beep! Route currently being replanned" blared incessantly from the GPS screen on the dashboard. Finally confident he was on the right track, Haohua hung up and explained that we were bringing a couple of wild-duck eggs for the bureau to inspect. The eggs were the product of a new "experiment" at the Po Lake farm, he told me. The wild ducks are fed a special concoction of Chinese medicines in addition to their normal feed. Wang Fenglian's idea is that the medicines in the feed will travel through the blood of the ducks and be retained within the eggs. When someone eats the eggs, they will also receive the benefits of the medical substances, which include improvements to the blood and heart, strengthening of resistance to viruses and bacteria, cures for high blood pressure, as well as remedies for dementia, back pains, and even aging skin. Inverting the logic of food-safety campaigns against pharmaceutical residues in eggs, Haohua explained that the benefits of good medicines would be transferred, residually, through the eggs. According to early experiments, Wang Fenglian later told me, two people who had eaten the eggs over six days had significant reductions in blood pressure.

Later that month, I traveled with Wang Fenglian and Haohua to the Second Quadrennial China International Forestry Expo, held in Yiwu, Zhejiang Province. Yiwu calls itself "the capital of small commodities." The Expo Center was brand-new. Massive balloons of blue, red, and yellow flew above, but rubble and dirt appeared only a few feet beyond the far side. We struggled to find a hotel and settled for a watery noodle soup for dinner. Yet Haohua said money flowed through Yiwu as quickly as the more-famous special economic zones. "They say there are more BMWs in Yiwu than in all of Germany!" Haohua exclaimed to a cab driver.

Inside the Expo Center's national pavilion, we found a section devoted to wild products. In the middle, ten mink coats hung on a rack. To one side, some kind of medicinal plant—long, brown, and bulbous—was displayed behind glass. On the other side, a man hawked snake oil. In the far corner, taxidermy birds perched, still as if pausing to watch us, colorful plumage above green artificial turf, with price tags as high as ten thousand RMB (around fifteen hundred U.S. dollars). In one small corner a sign confirmed the existence of a Wild Duck and Goose Zone. The small area was already filled by three or four exhibitors and their wares. We ran into the dean of the China Wild Animal Protection Committee, an organization of which Wang Fenglian is also a member: he is a vice dean of the China Goose and Duck Breeding Specialization Committee. Fenglian persuaded the man to make a phone call, and soon we were setting up promotional materials on a display case next to the other wild bird farms.

Back in the hotel, we sat on Fenglian's bed with the disassembled cardboard boxes, egg crates, and boxes of eggs around us. I folded the cardboard into boxes while Haohua carefully placed eggs into the crates. Then we put the crates in the finished boxes. When the expo opened to visitors the next day, Wang stood out in front of the counter, asking everyone who walked past whether they had high blood pressure. If they made any hint of response, he offered to give them a box of the miraculous eggs, no cost, in exchange for their contact information. Within fifteen minutes he had handed out two boxes of eggs and many more pamphlets, and had collected ten or so contacts in his small black notebook.

I returned one final time to visit the Wang family and the Po Lake farm in 2014. Fenglian and Haohua were still pursuing new enterprises and experiments. They now received government subsidies, for which they were required to increase the scale of the farm (*guimohua*), and had built several new farm buildings. Haohua was developing new connections for selling wild goose meat to restaurants in nearby provinces. He told me that he focused on marketing the distinctiveness of their wild breeds, which he compared to the brand-name coat he wore. They had even started selling farmed wild geese to a nature park for city children, where the park managers organize performances of geese in flight. As Haohua explained, "More and more people understand that ecology is more than mountains and water [*shanshui*]. In the water you need to have some fish swimming about, above the water something should be floating around, and in the air you need something flying." To say this ecology was artificial is an understatement. For the wild goose farmers, ecological relationships of wild and domestic could be better described as the objects of an ongoing experimental practice, constantly seeking to create distinctive forms and new values.[38]

FIELD DISPLACEMENTS

Laboratory ethnographies emphasize the role of experimental practice and material infrastructure in the production of new scientific knowledge.[39] The historian of science Hans-Jörg Rheinberger's concept of scientific displacement, which draws from historical and ethnographic studies of laboratories, provides a particularly sophisticated model. Rheinberger argues that "unprecedented events"—the surprising occurrences traditionally glossed as scientific discoveries—are in fact *engendered by* experimental practice and the laboratory apparatus: "They come as a surprise but nevertheless do not just so happen. They are made to happen through the inner workings of the experimental machinery for making the future."[40] Although made by experimental practice, they also "commit experimenters to completely changing the direction of their research activities." The paradoxical "research object,"[41] the object of scientific inquiry, is constructed by the experimental system yet remains irreducibly vague, embodying the unknown rather than the known and enabling the concerted creation of the unexpected. Rheinberger describes the production of scientific knowledge as a process of *displacement* rather than discovery, one in which the sciences "reshap[e] their agenda through their own action," but without foreknowledge of how their objects will take shape.[42]

When situated in the field, however, the scientific research object is also the object of other modes of creative practice that already inhabit the field site. This chapter articulates how these other modes of practice rework and transform scientific research objects, causing displacements that cannot be attributed to the inner workings of the experimental system. In flu research at Poyang Lake, the wild bird–domestic poultry interface constitutes a research object whose forms, the wildness and domesticity of the birds themselves, are at the same time being transformed by poultry breeders. I argue that displacements, the primary source of scientific change or discovery, can therefore result as much from an encounter with a poultry breeder's techniques as from the apparatus of the experimental system. In what I describe as "field displacements," scientific knowledge in the field develops through encounters with the outside of the experimental system.

In the design of their research object—the wild bird–domestic poultry interface—the NIH group presumed that wild and domestic birds are two distinct populations. The wild goose breeders, on the other hand, engaged with the wild-domestic distinction very differently: as a value differential that could be practically exploited to create something new. They were *breeding wildness*. When the NIH group encountered the swan goose breeders during field studies at Poyang Lake, they quickly saw the limits of their own concepts and developed new research objects. Yet this process differed from Rheinberger's model of laboratory displacement. For rather than deriving from the infrastructure and design of the experimental system itself—Rheinberger's "machine for making a future"[43]—the research object was displaced by poultry breeders whose breeding techniques and values reconstructed the interface of wild and domestic.

During their initial migratory-bird studies at Poyang Lake, Newman and other members of the NIH group stayed at the migratory-bird refuge in the town of Wucheng, surrounded on all sides by the Poyang wetland. Newman recalled that each day they set out to capture and mark wild birds in the refuge van: "To drive anywhere, from any point A to point B, you see lots of duck farming. We started talking to people, and then you get to know some of the local people, get to know some of the people at the wildlife reserve, and start talking, in those broader discussions, asking them about what was being raised, what kinds of species? So they started going into different species of ducks. And some of these were pretty unusual species to be raised, so we were wondering."

Later, after they asked their driver to stop by some farms, Newman encountered the swan goose breeder and his wild swan geese in the moment I describe at the beginning of the chapter. This encounter "led us over to farmed wild birds, [a] whole new level of interest," Newman told me, explaining that the encounter completely transformed the NIH group's understanding of "connectivity and interface between wild and domestic birds."

The shifting conceptual terms used by the NIH group to describe how connections form across the wild-domestic interface make visible the contours of this displacement. Before research at Poyang was begun, Newman coauthored the FAO technical manual introducing field research on wild birds and avian influenza. In a section explaining the hypothetical role of wild birds in influenza transmission, the authors point to the importance of what they call "'bridge' species": "Several bird groups without particularly strong ties to wetland habitats, but with a high tolerance for human-altered habitats, have also been known to become infected fatally from H5N1 [including crows, sparrows, mynas, and pigeons]. . . . These species may serve as links between wild birds in natural habitats and domestic poultry, acting as a 'bridge' in the transmission of AI viruses from poultry to wildlife or vice versa."[44]

Following the encounter with the wild swan goose breeders, the NIH group developed a new concept: "farmed wild birds." Clearly drawing on the earlier notion of bridge species, Newman explained to me in 2012 that farmed wild birds "could be the link between wild and domestic birds. They are the perfect intermediary. Because they look identical to their conspecifics, when they are foraging, a wild bird would come right up to them, because phenotypically they are the same. But then, they go home at night, and there are other poultry around at the farm. So there's your transmission!"

Yet despite resemblance to the earlier notion, the new concept subtly displaced the form of the NIH group's working object, the wild-domestic interface. In the original design of their pilot study, the NIH group understood the interface as a *spatial setting* in which contacts between wild and domestic bird populations took place, such as the fishponds described in Newman's presentation at the Living Lakes conference. A diagram of the original plan for the pilot study depicts white boxes marked "migratory birds" and "free-ranging ducks/geese" on either side of a blue oval identified as "Paddy rice fields/Natural wetlands/Fish ponds." The bridge species was an existing wild bird species that frequented such settings of interface, birds such as pigeons able to tolerate both natural and human-altered habitats.

With the concept of "farmed wild bird," on the other hand, the researchers transposed the conceptual boundary between wild and domestic from a spatial setting or habitat to the bird itself. In doing so they drew attention to the breeding practices that cultivate birds able to double as either wild or domestic, practices that *internalize* the wild-domestic interface within the farmed wild bird. The subsequent research projects the NIH group conducted at Poyang Lake made the significance of this displacement clear. Following the discovery of the farmed wild bird, the NIH group focused inquiry on the human practices responsible for breeding wildness: they counted and mapped the households that farmed wild birds, conducted surveys to understand vaccination regimes, and followed the market chains along which farmed wild birds were traded. And when the NIH group updated its diagram of the wild-domestic interface to include farmed wild birds, this new vector of human agency was also added, a new white text box containing the words *production, market, trade, transport systems, vaccine, movement control, culture, behavior.*[45]

The field has not entirely replaced the lab in influenza research; indeed, laboratory analysis of viral samples remains an important component of flu research at Poyang Lake. Rather, the movement into the pandemic epicenter displaces the predominance of laboratory *practice* as a model for understanding the process through which new scientific knowledge is made. For the classic laboratory ethnographies, the lab was a tactical site where science could be studied as a cultural practice, thereby calling into question the importance of theoretical structures and mental cognition as sources of scientific knowledge and change. Yet this focus on experimental practice, the significance of which is so evident inside the laboratory, has obscured from view the more-variable trajectories of scientific discovery.

Of course, the defining features of laboratory practice could be found in the initial setup of the migration study at Poyang Lake. Transponders attached to wild birds sent signals to orbiting satellites, transforming migratory movements into detached "traces," marked with spatial and temporal coordinates, and available for scientific manipulation and analysis back in laboratory centers.[46] In the end, however, the traces detached from the flights of birds did not displace the NIH group's research object; an encounter with poultry breeders did.

When Newman first told me about farmed wild birds, he laughed and recalled how he had posed as an American poultry buyer on his first trip to a Poyang Lake farm. Whether or not he was fooled, the breeder went along with the performance, asking which seaports would be most convenient for shipments to the United States. That evening, he sent a whole, fresh-killed swan goose to Newman's hotel. The insights about the farming of wild birds that shifted the NIH group's research objects came not from an extension of his laboratory to the lake but rather from Newman's momentary abandonment of the subject position of scientific expert. By taking on the pose of the buyer, Newman came to understand the concept of wildness guiding the practice of wild goose breeding, an understanding that the "inner workings" of his experimental system could not provide.[47]

Moving influenza research into the pandemic epicenter constructs research objects on "working landscapes," sites of already ongoing labor and production. This co-labor on the same sites can cause unexpected field displacements to scientific research objects. When both scientists and breeders work on the same birds, breeding techniques become as important as experimental design for the production of novelty and surprise that lies at the heart of scientific change. But recognizing these field displacements, and thereby incorporating them in the trajectory of knowledge production, requires putting aside, at least for a moment, the laboratory practices of detachment and purification. In his encounter with the breeder, Newman relied on techniques more familiar to the human than the natural sciences.[48] What had once looked to him like any other wild swan goose, *Anser cygnoides*, he now saw as a product of human practices. Playing the part of a participant in the breeders' world of food markets and gourmet tastes, Newman strove to understand the ideas and values driving the cultivation of wildness.

Influenza research at Poyang Lake describes an anthropological arc of sorts, one in which knowledge of natural objects first passes through an understanding of human engagements with the natural environment. Many studies have shown that the sciences today are forming new relations with society through patients' organizations and bioethics regimes, transforming how knowledge is made in the process. Here I argue that the changing sites and objects of contemporary influenza research are shifting the epistemological relation of the sciences to nature as scientists in the field come to see natural sites as human artifacts. Laboratory ethnographies expose the material infrastructure and scientific labor required to construct the spaces where scientists encounter nature and take its measure. The field displacement of influenza research at Poyang Lake reflects a different epistemological question: How do scientists account for the practical engagements, such as poultry breeding, that creatively transform the natural sites where field experiments are undertaken?

REGULATORY WILDS

One day I rode the small fish-boat ferry from the refuge base in Wucheng to a small island across the river where I met one of the other wild goose farmers in the region. Dressed in a bright pink shirt, Ye picked me up at the ferry landing on his motorcycle, and we rode through town to his farm, perched at the very edge of the lake. A series of low sheds for the geese and a freshwater pond framed his house. In the past the village had used the pond for a collective duck farm, but it was no longer operational when Ye contracted for it and surrounding lands in 2003. When I visited, about three thousand swan geese swam in the pond.

As we walked around the pond, Ye led us up a small embankment, and there was the lake. Grass sloped down to the shore. Beyond, water stretched as far as the eye could see. Ye said the geese sometimes fly as far as three kilometers away to eat the lake grasses. Pointing at the embankment, he showed me the high-water line and told me that the previous year the lake had even breached the embankment, flooding his entire farm. As I later learned, the flood had been costly—of the ten thousand swan geese they were raising at the time, more than half had died. Ye explained that the water submerged all of the grass, so the birds had nothing to eat. Ye's wife also added that pesticides from neighboring crop fields flowed into their farm, sickening their birds: "it was like an epidemic." Because of these risks, and because of the instability of the market, Ye now raised a thousand sheep and several hundred cows in addition to the swan geese. He described this as a strategy of diversification: risk should be spread around among different breeds, he explained, because if one of the breeds completely fails for some reason, then the others can make up some of the losses.

I asked him whether he thought of the geese as wild (*yesheng*) or domestic poultry (*jiaqin*). "In China, we call this 'special type husbandry' (*tezhong yangzhi*)," Ye replied. "In fact, they belong to forestry (State Forestry Administration), not agriculture (Ministry of Agriculture)." In order to raise swan geese, he first needed to apply for a license from the SFA. The SFA sent a team to inspect his farm, to see that the conditions are good, and would not cause harm to the environment. But Ye acknowledged that SFA was not very experienced with animal husbandry. "They still mostly deal with forests." The SFA did hold a special conference for wild goose farmers, in which Ye got to meet other farmers from across Jiangxi and Zhejiang provinces. Even at the conference, though, a Forestry official admitted to Ye that the SFA did not know much about raising swan geese. "Are they even good to eat?" the official joked.

Ye found the classification especially irritating because it meant he could not apply to the Ministry of Agriculture (MOA) for subsidies offered to livestock farmers. "I don't have to pay their licensing fees, but I don't get the subsidies either." However, this classificatory logic also had significant implications for the ecology of influenza, because of how it effected the implementation of vaccination for H5N1 avian influenza. In 2005, China's Ministry of Agriculture had begun implementing a program for universal mandatory H5N1 vaccination. The program literally called for immunization of 100 percent of poultry (*jiaqin*)—the domestic chickens, ducks, and geese regulated under the MOA. Wild geese, by contrast, despite the fact that biologically they closely resemble domestic geese, are not administered by MOA, and therefore do not fall under the immunization plan.

When I asked Ye if he had suffered any outbreaks of disease in the geese, he explained that he vaccinated all the geese himself, and—combined with the good environment and remote location—this kept disease away. But when, in response to experimental displacement, FAO began to shift their research projects to look at wild bird farming in the Poyang Lake region, they found something troubling about vaccination practices. Based on a survey of wild bird farms around the lake, wild bird biologist Changqing Ding found that vaccination for H5N1 avian influenza was inconsistent and irregular. Although most farmers did conduct some form of vaccination, the number of shots, intervals between doses, and timing in the birds' life-cycles all varied widely, suggesting that immunization was probably for the most part ineffective.[49] In addition to the risks created by the creative remaking of the wild-domestic interface, the regulatory anomaly of the farmed wild bird created risks of its own. The ecology of influenza, rerouted along the changing working landscapes of rural production, was also reshaped in unexpected ways by the territories of regulatory governance.

In part I of this book I argued that the logic of the pandemic epicenter produced a spatial displacement toward the epicenter. Because the epicenter is located within a *working landscape*, part II showed that scientific inquiry at the epicenter moved along a pathway toward the outsides of the laboratory venue and the expert subject. In contrast to laboratory detachment, I have drawn attention to these centrifugal displacements, explications of the externalities of biosecurity models or experimental systems. In part III, I examine the consequences of the pandemic epicenter's location in geopolitical space or, more precisely, its location within China's national territory. How does the displacement of scientific attention to the epicenter intersect with the sovereign authority of the nation-state? What reconfigurations of geopolitical space take place when global health goes into the epicenter?

PART III

TERRITORY

CHAPTER FIVE
AFFINITY AND ACCESS

Just after Chinese New Year in 2011, as Beijing was slowly returning to work following the month-long holiday, I found Vincent Martin in the Emergency Center, sharp lines of frustration already beginning to crease the tan acquired during a beach vacation in Thailand. He had returned from Phuket to a regional outbreak of foot-and-mouth disease that, he explained, China had so far refused to report. I turned the conversation toward influenza, and Martin kept on the theme of China's lack of "transparency." He complained that for over a year not a single avian influenza outbreak had been reported by the Ministry of Agriculture to international agencies. Martin believed China was "in a bind": if they report any outbreaks now, it would seem like the situation is getting worse, not better. "Couldn't you just report one or two?" he asked rhetorically, a thin grin easing the lines on his face. "There have been outbreaks reported in Cambodia, Vietnam, Korea, Japan, but none in China?"

Martin's denunciation of China's failure to report outbreaks of disease reflects an enduring controversy over China's relationship to international health organizations within an emerging regime of global health.[1] Recently rooted in the epidemic of severe acute respiratory syndrome (SARS), especially China's delayed acknowledgment of the explosive spread of the previously unknown atypical pneumonia, the affair extends to disputes over the sharing of influenza virus samples and the exchange of health information. In the case of the avian influenza outbreak, global health agencies have turned their suspicion away from China's Ministry of Health—increasingly considered a cooperative global partner—and toward the Ministry of Agriculture (MOA). For example, complaints about failures to share samples and genome sequences of flu viruses flared up in 2005 and 2006. "We really don't know how many strains of bird flu there are in China," the WHO's coordinator of epidemic response in China publicly announced in November 2006, "because we have limited amounts of information shared with us by the Ministry of Agriculture and the virus samples we have asked for have not been shared."[2]

Most accounts have described China's failures to report outbreaks, or refusals to share virus samples, as "assertions" of national sovereignty in conflict with global health norms of open-access research and transparent surveillance.[3] China's disputes with global health agencies are compared with affairs and incidents elsewhere, especially Indonesia's well-known refusal to share influenza samples. Although scholars are divided on how to appraise the justice of China's actions, they have agreed on the basic structure of these disputes: a static normative opposition between global circulation of health knowledge and the proprietary claims of nation-state authority.[4] Complaints by international health civil servants, like Martin's above, sometimes seem to share this perspective that the management of public health information puts the Chinese state at odds with global health norms.

However, I found that international flu experts engaged in diverse transactions with Chinese counterparts despite, and alongside of, their frequent criticism of China's lack of transparency. Many of these transactions moved samples, sequences, or other forms of information across boundaries declared in other moments to be impermeable. By drawing attention to concrete situations, describing actual practices, and expanding the temporal scale of analysis, polemic disputes and critical accusations no longer appear as the signs of a static conflict between China and global health organizations. Rather, these disputes appear as one part of how affiliations are given concrete form and structure, building relationships that *regulate* exchanges of viral materials and information.[5]

These borderlands of health knowledge and biological exchange, involving multilateral development agencies, nongovernmental organizations, national governments, and others, have been described as a contested domain of "global health diplomacy."[6] In this chapter I provide an account of the everyday practices of international civil servants who could be called global health diplomats. Rather than presuming a static opposition between national sovereignty and global health, I situate disputes within the actual, uncertain, and dynamic transnational relationships of affinity and exchange. I show how the "international" and the "national," far from stable and opposed entities, come to be marked and differentiated through a process of transnational affiliation.

Geopolitical space has garnered significant attention in studies of humanitarian global health interventions, which have highlighted how humanitarian interventions may constitute "migrant sovereignties" and "spaces of exception" that dispossess certain sites or bodies from national territorial space.[7] In a study of Doctors without Borders, Peter Redfield provides a nuanced account of how humanitarian actors confront ethical and political dilemmas regarding the spatial juxtapositions of national territory and transnational humanitarianism. But the geopolitical spaces constituted by regimes of global health security, focused on the control of emerging infectious diseases, thus far remain less well mapped.[8]

By following Vincent Martin's global health diplomacy, this chapter traces an ethical journey into the pandemic epicenter.[9] I found that Martin adopted two distinct strategies of access to the epicenter. On the one hand, he adopted a strategy of *affinity*—that is, he cultivated a network of associations with Chinese state agencies and actors in order to exchange and communicate. Like the matrimonial strategies described by Pierre Bourdieu, Martin cultivated affinities in order to set people, information, and material goods in motion.[10] Affinity was often an indirect strategy of access as it passed through the detour of friendships in order to subsequently access biological materials or key information.

Yet in other moments, Martin adopted another strategy altogether: he built equipment that obviated any need for affinity and exchange by moving to another plane of reference, a strategy that I call *stratification*. As we will see, this often entailed supplanting political space with representations of ecological space, surpassing the geopolitical strata of territory with concepts such as ecosystem or habitat that occupy different scales and spatial forms. In the discussion that opened this chapter, for example, Martin's reference to the known occurrence of outbreaks in the region (an ecological unit) juxtaposed with the absence of reported outbreaks in China (a geopolitical unit) was intended to expose the obstruction of scientific observation by politics. In such moments, Martin's claim that an emergency situation justified global health interventions into China resembled the logic of exception deployed by humanitarians. Unlike humanitarian interventions, however, Martin did not figure planetary humanity as an ethical imperative for bypassing nation-state sovereignty. Rather, he aimed for a technical demonstration of the insignificance of national scale or territorial boundary in the ecology of pandemic emergence.

By the middle of 2011, China's Ministry of Agriculture had issued several statements heralding a remarkable fact: there had been no outbreaks of HPAI H5N1 for nearly two years. However, as Martin's skeptical questions make clear, not everyone believed that the lack of outbreak reports indicated an absence of disease. For one thing, routine avian influenza surveillance programs conducted by the MOA continued to report the isolation of viruses, as well as serological evidence of infected birds, despite the lack of sick birds or outbreaks. Martin and others began to raise concerns about the silent circulation of H5N1 viruses throughout the country, which could still cause human infections and drive the emergence of new viruses. Almost everyone agreed that the dramatic reduction in H5N1 outbreaks, but failure to completely eradicate the H5N1 virus, was a troubling and unexpected consequence of one state initiative: the universal immunization of poultry.

In chapter 3 I noted that the initial FAO strategy for responding to HPAI H5N1 included three prongs: culling, biosecurity development, and vaccination. The FAO was initially reluctant to promote vaccination: the method had been applied only in two outbreaks of highly pathogenic avian influenza in Mexico and Pakistan, both during the 1990s. Highly pathogenic avian influenza viruses had been eradicated using culling, movement, and trade controls in all other outbreaks since the 1950s. However, as FAO documents show, the organization quickly acknowledged that vaccination would be an important "ancillary tool" because H5N1 was becoming endemic across the Southeast Asian region.

China's effort to develop a vaccine for highly pathogenic avian influenza began in the 1990s, around the time of the Mexico and Pakistan outbreaks. More or less since its inception, the vaccine program has been led by Chen Hualan, currently the director of the National Key Laboratory for Animal Influenza at the Harbin Veterinary Research Institute (HVRI, in Harbin City, Helongjiang Province), itself an institute within the umbrella of the Chinese Academy of Agricultural Sciences (CAAS). At the time, Chen was a doctoral student in preventive veterinary medicine at the HVRI, under supervision of the institute's director, Yu Kangzhen. Yu is credited with initiating the HPAI research program at Harbin, beginning with the import of several influenza strains (representing each of the antigens H1–H15) from the Central Veterinary Laboratory, Weybridge, United Kingdom. The strains were given to HVRI by Dennis Alexander, who led the Weybridge lab.[11] Yu assigned Chen and four other colleagues to work on influenza and specifically on the development of a vaccine. In the first years of her research, Chen recollects that "we couldn't do any basic research because we didn't have too many strains at that time."[12] Moreover, although things were better than in the 1980s, equipment remained scarce: for instance, she would have to apply two to three weeks in advance in order to run material in a PCR machine.[13]

After the Harbin lab isolated the H5N1 highly pathogenic avian influenza virus from a goose in Guangdong Province in 1996, the national government began to support the development of an H5 vaccine for use in poultry.[14] The program gained urgent top-level encouragement after the related, but genetically distinct H5N1 virus emerged in Hong Kong in 1997, causing widespread poultry outbreaks on farms; causing seventeen human cases, including five deaths; and prompting predictions that a pandemic could be imminent. In 1999, in the midst of this vaccine development work, Chen applied to become a postdoctoral researcher with Kanta Subbarao, chief of the Molecular Genetics Section of the U.S. Centers for Disease Control. For Subbarao and the CDC, hiring Chen was a useful opportunity to quickly gain inside knowledge on the emerging H5 viruses in China and Hong Kong. Chen, on the other hand, gained access to the most advanced techniques of influenza research and vaccine development.[15]

Chen's most important lesson at the CDC was training in an emerging technique known as plasmid-based reverse genetics. At a broad level, reverse genetics refers to the practice of creating genetic mutants and then examining what phenotypes appear, the "reverse" of classical genetics approaches that attempted to find the genetic basis for a particular, known phenotypic trait. Since the 1970s, influenza vaccine production has involved the laboratory creation of reassortant viruses that combined antigenic properties with desirable attributes sourced from other strains. Usually, this involved recombining the high growth rate from a standard laboratory strain (e.g., H1N1 AO/PR/8), necessary for industrial manufacturing, with the surface glycoproteins of the antigenically targeted currently circulating strain. However, the production of reassortants by co-infection is "cumbersome and time-consuming" because each co-infection event can "theoretically result in the generation of . . . 254 different progeny viruses," each of which needs to be carefully screened in order to select a suitable seed virus.[16] Plasmid-based reverse genetics, an innovative technique first developed in labs associated with Peter Palese and Robert Webster in the late 1990s, promised a much faster, and more fine-tuned, process of artificial recombination and strain selection.[17]

Chen returned to Harbin in 2002. Soon after, the laboratory developed an inactivated vaccine built from a low-pathogenic H5N2 virus, one among the low-path viruses originally sourced from Weybridge. In 2003 the government approved the vaccine for use, first in Guangdong Province farms that raised chickens exclusively for export to Hong Kong and Macau, and then later for more widespread use. Almost immediately after the vaccine was approved, the H5N1 virus "reemerged" in Hong Kong and spread throughout Southeast Asia.

In response, China's MOA began to conduct "buffer zone vaccination" in and around outbreak sites using the inactivated H5N2 vaccine.[18] In 2004, 2.5 billion vaccine doses were used.[19] However, as Chen Hualan noted in a review of the vaccine program, the vaccine was "not ideal" because the "seed virus exhibited antigenic diversity with the prevalent H5N1 strains circulating in China at that time."[20] In addition, the seed virus did not grow well in eggs, making it a poor choice for large-scale industrial vaccine production.[21] In order to produce an antigenically better match, Chen and the Harbin lab began building a new vaccine from scratch in the laboratory. Using plasmid-based reverse genetics techniques, they created a reassortant seed virus that combined the antigen-binding genes (HA and NA) from the 1996 Guangdong goose H5N1 virus with internal genes taken from a high-growth influenza A virus. The technique also eliminated the genes for high virulence, making the vaccines safer to use. As a published reflection from one of the lab members points out, plasmid-based reverse genetics "solved all of the problems we had faced in the research on avian influenza vaccines."[22] Chen and her team named this vaccine Re-1, meaning "first reassortant vaccine."

China began using the Re-1 vaccine concurrently with the inactivated H5N2 vaccine in 2004 for buffer zone vaccination. Production was supplied by a limited number of government-approved pharmaceutical companies, including the Harbin Weike Biotechnology Development Company, a "state-owned" enterprise partially owned by the Harbin Veterinary Research Institute itself.[23] Although the vaccine was effective in controlling outbreaks, it was soon clear that vaccinating in response to outbreaks was not preventing the emergence of new outbreaks. Chen remarks in a published reflection that in 2005, "epidemiological studies indicated that all the prior outbreaks had occurred in farms that did not vaccinate or vaccinated with unqualified vaccines."[24] Under pressure from the World Health Organization to control outbreaks before they seeded a pandemic, China took a dramatic and unprecedented step. On November 11 the Ministry of Agriculture issued a new policy standard that stipulated mandatory universal immunization of *all poultry* in China against H5 highly pathogenic avian influenza. By universal immunization, the new policy literally required the vaccination of "100 percent" of poultry (including chickens, ducks, and geese)—estimated to be around 14 billion birds.[25]

Although supporting vaccination, officials from international agencies such as the FAO immediately expressed skepticism over the scale of the immunization policy: how to get vaccines to so many birds across such a large territory, particularly considering that many are "loose" (not raised in enclosed housing) on so-called backyard farms?[26] A closer look at the policy shows that the target of "100 percent" was to be achieved by dividing poultry farms into "routine" and "scattered poultry" (*sanyang jiaqin*) types, a distinction linked to broader distinctions in China's agricultural development programs that I discussed in chapter 3.[27] The distinction between "routine" and "sanyang" immunization reflected two completely different mechanisms addressed to two different types of farm. On the one hand, so-called scale farms (*guimo yangzhichang*) were required to conduct routine vaccination using their own veterinary staff and equipment. Routine immunization protocols required vaccinating all birds when they were fourteen days old, then three weeks later injecting a booster vaccine, and vaccinating again every six months.[28] This immunization would be recorded by county-level veterinary agencies and in some cases verified by postvaccination serology. By contrast, government agencies directly conducted immunization of all sanyang poultry in annual campaigns each spring and fall.[29]

The impact of what I am calling the geopolitical strata of the pandemic epicenter becomes clear here. In the vaccine program, China linked biotechnological innovation with industrial manufacturing and a mass campaign intervention model with roots in the Mao era.[30] The policy, which made *universal* immunization both mandatory and free of charge, resembles emergency health campaigns to control schistosomiasis and SARS, but with one crucial difference: the campaigns shifted the object of national intervention from humans to birds. By imposing a poultry immunization program at the scale of the national territory, China extended the domain of national biosovereignty over the living bodies and populations of chickens, ducks, and geese.

In doing so, this national intervention produced significant effects on the landscape and ecology of the epicenter. Widespread, or near universal, vaccination transformed the ecology of the H5N1 epidemic in a number of possible ways. Most controversially, some Hong Kong scientists speculated that vaccination in China may have enabled the emergence of new strains of influenza through a process of escape and evolutionary selection.[31] Representatives from China's Ministry of Agriculture and the Foreign Ministry denied the accusation and claimed that surveillance programs showed "no distinct changes in biological characteristics" of the virus.[32] Many observers suggested that China's universal vaccination program had not eliminated the circulation of the virus but only rendered it invisible. Whatever the case, China's poultry immunization program undoubtedly remade the biological immunity of avian populations, shifting the host ecology of influenza at a national scale.

The global health response to the HPAI H5N1 outbreak, and in particular FAO's programs to contain the emerging virus "at source," directly confronted the overlay of China's national biopolitics upon the cultural landscapes and viral ecologies of the pandemic epicenter. In the rest of the chapter I follow Vincent Martin's journey into the epicenter, focusing on his role in establishing and directing the FAO Emergency Center in China. I trace how the Emergency Center's pathway intersected with the geopolitical strata of national biosovereignty, leading to displacements of global health diplomacy toward new networks of affinity.

THE EMERGENCY CENTER

Immediately following the reemergence of HPAI H5N1 in 2003, as the FAO began to develop a strategy for controlling pandemic influenza "at source," Martin wrote a two-page concept note outlining the design of an Emergency Center for Transboundary Animal Diseases (ECTAD) at the FAO headquarters in Rome. As Martin described to me, the concept note emphasized new pathways of collaboration: "Within FAO we also realized that the magnitude of the problem required a different approach, more coordination, a multidisciplinary approach, having a transversal approach to this problem, and we were not organized this way within FAO; we had our different departments, and services, working on animal health, on animal production, or economics, and we realized we needed a platform of experts from different horizons to work on this crisis."

The logic of emergency inscribed new lines across the institutional form of the FAO, bringing together specialists from different disciplines to analyze information about the emerging outbreak. Yet the validation and interpretation of reports from national governments remained a challenge. As one ECTAD staff member reported, "We quickly reach the limit of our system. We need expertise in the corridor to recognise what is going on." As this staff member complained about one (unnamed) country, "If they report, it's because everyone already knows. The key question, when it gets serious, is the high level of expertise we need in the corridor. It is more and more difficult to find good people."[33] As a result, the FAO began to shift experimental systems and biosecurity interventions toward the epicenter, juxtaposing the *transversal* geometry of collaboration with the long-standing international orderings of the UN organization.

At first, the response remained framed within the FAO policy known as technical cooperation. Established in 1976, the Technical Cooperation Program (TCP) shaped largely top-down transfers of technical knowledge and aimed for what historian Amy Staples calls "the transfer of expertise" from developed to developing nations.[34] According to the *TCP Manual*, "The Technical Cooperation Programme . . . aims to provide FAO's technical expertise to its Member countries through targeted, short term, catalytic projects. These projects address technical problems in the field of agriculture, fisheries, forestry and rural livelihood that prevent Member countries, either individually or collectively, from implementing their development programmes." More specifically, "TCP projects aim to fill the critical technical gaps by providing technical inputs that are not available locally, or that project beneficiaries cannot access through their own means, or through local support systems."[35]

This TCP model is rooted in the logic of technical internationalism that guides the FAO's developmental work, largely in common with other Bretton Woods and UN agencies. Indeed, the first director-general, John Boyd Orr, once described the FAO as "an international extension agency." He argued that "the resources and powers entrusted to FAO are woefully limited in relation to [its] far-reaching objectives. It cannot order particular policies to be adopted; it can only advise, educate, and persuade. It cannot embark on the executive functions of purchase and procurement in order to stimulate output and equalize distribution; it can only recommend, demonstrate, and discuss."[36] In part because of this advisory, rather than executive, function, the FAO often articulated the international as a technical and humanitarian domain distinct from politics.[37]

In the initial response to the reemergence of HPAI H5N1, the FAO established TCP agreements with many countries in the region. TCPs are paid for out of FAO general funds, rather than by funds from donor countries, which enables a more rapid response to events. However, a national government must make a formal request for technical assistance in order for a TCP to be authorized. In this regard, most countries made requests for laboratory training or laboratory equipment, but China was different. As Martin explained to me in conversation,

> [the FAO] sent a different kind of expert [to China]. . . . For them, strengthening laboratory capacity was not a big issue. So we knew there was a national reference laboratory in Harbin, we knew they had huge capacity to produce vaccines and different types of vaccine according to the strains, so this is not where we put the emphasis. We did put the emphasis more

on surveillance and epidemiology, and this is how we realized, during this project, but it was also my observation, the weakest link was really in epidemiology. They were very capable of possibly, well, diagnosing the disease, typing the type of virus, sequencing the virus, producing the vaccine, and having a sound control strategy. All that was kind of OK, but the missing link was really on the capacity to understand the big picture, of the disease, the ecology of the disease, where it is, how does it spread, where are the most high-risk areas, how are we to better target our vaccination strategy or our surveillance system in live-bird markets.

Most of the initial FAO TCP "emergency assistance" donation of $387,097 to China was earmarked to "gradually strengthen national capacity in epidemiological investigations," with a small sum given for laboratory diagnostics.[38] The FAO hired one national veterinarian, Cai Haifeng, a career employee of the China Animal Health and Epidemiology Center (CAHEC, formerly known as the Animal Quarantine Institute), to begin early epidemiological work on avian influenza in China.

Yet Martin remained frustrated by the top-down vertical lines of technical cooperation. By 2005, he began "pushing for having an ECTAD office also in China because I thought that it was meaningless to work in all the surrounding countries, trying to curb the spread of disease, while the epicenter—if we may say so—was in China in a way and it was not good just to have remote collaborations with them." Martin explained that the ongoing TCP project aided his request for clearance from the Ministry of Agriculture. However, because the ECTAD China office would entail the full-time presence of an "international expert" (Martin himself), he still worked for more than a year to successfully get this clearance. As he told me, China "was also quite difficult to get in, to have such a close relationship . . . as we had with the other countries."

The displacement of the Emergency Center to China began to crisscross the FAO's structure of technical internationalism with new lines of collaboration. Martin located the Emergency Center office in Sanlitun, Beijing's international diplomatic district, at some distance from the FAO Representation in China, which has offices in the commercial center of Guomao. The Emergency Center was on the fifteenth floor of a twin high-rise office tower, surrounded by quiet, tree-lined streets that mostly contained older diplomatic compounds. The United Nations, the World Health Organization, and countless international NGOs had offices nearby. Martin explained to me that the ECTAD national and regional offices are mostly independent from the FAO country offices, such as the FAO representation in China. Instead, they are all linked directly to the central Rome headquarters and the ECTAD office there. Moreover, their funding comes almost exclusively from direct, project-based donations by donor countries rather than from FAO general funds. The Emergency Center office in Beijing, for instance, is funded almost exclusively by donations from the United States Agency for International Development (USAID). This independence enabled significant flexibility in its projects and programs, which skirted around some of the diplomatic formalism built into the FAO's international structure.

However, the FAO's technical internationalism remained powerful, creating a very different context for collaboration compared with university research or NGO-led health interventions. Traditionally, the FAO has always stressed the "international" character of the staff at the headquarters in Rome. Andre Mayer, an early chair of the FAO Executive Committee, recalled that after one year "of seeking in turn advice from Chinese or Hindu, New Zealander or South African, South or North American, or from a European colleague," he came to realize "that their hearts and minds are in agreement on problems which concern all of them and that they think only of solving them for the common welfare, [so] one can look upon the future with hope."[39] As early as 1952, the FAO employed technical experts from forty-one countries.[40]

At the same time, the offices located inside member states, including the emergency centers, maintain a rigid institutional distinction between "international" and "national" staff. Aside from Martin, every staff member of the Beijing Emergency Center is a Chinese citizen and is formally referred to as "national" staff. When the Emergency Center posted hiring calls for such positions, it always specified the attribute "national" in the job title, indicating that they sought a citizen of the host member state. Cai Haifeng, for instance, held the title of "national technical advisor."

The FAO ECTAD office often invites specialists in a variety of fields to China: coming from as far away as the United States and Australia or as near as Thailand. The FAO refers to these visits as "missions" (including assessments, lectures, or training courses). The specialists are always identified as "international experts." In my own consulting contracts with the FAO, I have been referred to as an "anthropologist-international expert" and "international consultant in medical anthropology." Sometimes the need to insert "international" could produce awkward-sounding sentences. Under the headline "International Health Experts Visiting China," a 2009 article from *China HPAI Highlights* (the newsletter published by the Emergency Center) reports that "three animal health international experts" visited the Emergency Center in Beijing. These visitors are experts in the field of animal health, but their expertise is also qualified as having an *international* source and character, despite the strange phrasing that results.

The Emergency Center office layout reiterated this vision of the relationship between national and international in its architectural plan. The "national" staff all work at computers in small, semi-enclosed cubicles within the main office. On the side of the room nearest the entrance, Cai Haifeng, the most senior national staff member, works inside a glass-enclosed office. Next to this is another glass room with a conference table, where meetings with visitors typically take place. On the far end of the main room, a small opening leads into Martin's corner office, separated from the rest of the space by a short baffle passage rather than a door. This passageway gives the international senior technical coordinator a physical independence from the rest of the office—both sound and sight are blocked—while maintaining an opening for communication. Whenever I was in Martin's office, staff members would enter with a question or a piece of news, or Martin would pass outward to pick up a printout or to follow up on an email. The glass walls of the conference room expressed an ideal of transparency and public oversight: discussions inside could be seen, though not heard, from any point of the main office space. The senior technical coordinator's office, by contrast, both marked a border between international and national status and facilitated movement across it. More broadly, although the Emergency Center institutionally inscribed a boundary between "international" and "national," as we will see this boundary was less an impermeable border (or even a solid, but transparent, glass wall) than it was a synapse or passage: the distinction between international and national created transboundary openings that moved expertise *into* the epicenter and virus samples, genome sequences, and research data *out*.

DETOUR AND ACCESS

Running along the header of every document issued by the Beijing Emergency Center is a remarkable logo. On a red ground that resembles a billowing Chinese flag, a chicken and a duck stand face-to-face, drawn in white silhouette. In their hand-like wings, the birds hold glasses of red wine aloft and barely touching as if only a moment ago each had urged their counterpart to drink a celebratory toast (see figure 5.1).

The logo, which Martin crafted himself with the help of a design company, expresses normative ambivalence toward relationships of communion and communication. To anyone working amid the bird-flu crisis, the clink-clink of glasses and sharing of wine by duck and chicken contains an unsettling irony:

Emergency Center for Transboundary Animal Diseases (ECTAD) in CHINA

Emergency control of HPAI
OSRO/RAS/604/USA

FIGURE 5.1. The logo of the Emergency Center.

after all, research suggests that contact between ducks and chickens plays a significant role in the emergence and persistence of highly pathogenic avian influenza viruses.

Yet on another plane, the logo also evokes Martin's ideal of collaboration and his vision for the ECTAD office in Beijing. At one of our first meetings, Martin told me that he faced two challenges to collaboration, challenges that at the time seemed to me incredibly heterogeneous: veterinarians don't want to work with medical doctors, and China doesn't want to share influenza sequence data with the FAO. As the logo suggests, Martin's approach to China's unwillingness to share sequence data was not public outrage or critique but rather the building of relationships across difference, reframing international work in the terms of interdisciplinary collaboration. In this reading the wine shared by the two birds stands for the shared work across disciplines or territorial borders. And in the many ECTAD meetings or events I attended in Beijing, often jointly organized with China's Ministry of Agriculture, this metaphor approached the literal, when Martin or the Chinese hosts inevitably stood and asked us, in the Chinese phrase, to *ganbei* (empty our cups).

Drinking alcohol, and especially drinking *baijiu* (a white liquor usually made from sorghum), is an important component of banquets in China's business and bureaucratic circles. Banquets are sites for cultivating *ganqing*, or sentiment relations, among participants, and they help to extend and strengthen *guanxi* relationship networks. Banquets are efficacious but not strictly instrumental: the reciprocal exchange of toasts produces new spaces of shared sentiment that can persist long after the warm glow of the alcohol fades away. Although most popular culture references to banquets and baijiu focus on China's business worlds, banquets are also common within professional and bureaucratic sectors such as medicine and public health.[41]

Discussions of guanxi and banquets are ubiquitous in the anthropological literature. They are usually depicted as an archetypical Chinese cultural form, even if one that has increased in importance during the post-Mao era of rapid economic growth.[42] Therefore, little attention has been given to the role of banqueting in *transnational* spaces, beyond a few anthropologists reflecting anecdotally about their own experiences as a foreign guest. Yet it soon became clear to me that banquets were an important part of how the Emergency Center got its work done in China, even if the actors could in no way be said to share a "local moral world" in common.[43] By contrast, for Martin banqueting and other guanxi-like affinity practices were precisely about working with difference, or strategically engaging with how he understood things got done in China. Martin told me, for example, that the Emergency Center logo was partly an allusion to what he called "baijiu parties" that he had conducted as he sought to get Chinese government approval for various influenza research projects. Banquets, or baijiu parties, helped break down boundaries of difference, including political boundaries, in order to make new forms of collaboration possible. Through these liquor-filled banquets, the staff from the Emergency Center could transform the transnational relations of communication, increasing the flow of viral samples or sequences from China to the international agency, for instance, without resorting to the unconditional demand for transparency.

In 2009 Martin organized one of the Emergency Center's first projects in China: a research trip to Poyang Lake to collect samples of H5N1 virus. As I discussed in previous chapters, Poyang Lake had recently attracted attention from global health agencies as a possible pandemic epicenter. When Martin put forward his proposal for a sampling study at Poyang, however, the Ministry of Agriculture flatly refused the first request for research clearance. Still, Martin was not deterred. "You talk to a few people, drink a little baijiu, and all of a sudden you're among the birds at the lake," he told me. A report later published as the first issue of *China HPAI Highlights*, the newsletter of the Emergency Center, declared the mission a success, particularly highlighting evidence of collaboration:

> The project, funded by Sweden, collected samples from domestic ducks
> farmed near Poyang Lake and was very successful in bringing together

mission members with Jiangxi Province veterinary services and local veterinary authorities to work collaboratively in the field. All parties were ultimately able to agree on the methodology; and, proper sample collection techniques were demonstrated and used in executing the project. . . . The working relationships formed as a result of this project are very valuable to FAO's continued work in China to combat HPAI. This field mission can be considered a success on many fronts, and we believe that similar field data collection efforts in the future are more likely to be accepted and successful as a result of this project. Approximately 60 epidemiologists and laboratory staff were trained during this mission.[44]

The Poyang Lake project reflects a mutation in the form of the FAO's model of "technical assistance" and, as a result, an important change to how the FAO structured relations between the "international" and the "national." As I noted above, technical assistance often takes the form of training programs, aiming for the "replication of experts."[45] The Poyang Lake project, for example, involved the training of sixty epidemiologists and laboratory staff. Yet the hierarchical and pedagogical form of technical assistance seems to stand in tension with the language of "work[ing] collaboratively in the field" used in the newsletter to discuss the project. Through this juxtaposition of hierarchical and horizontal relationships, the Poyang Lake project reconfigured "technical assistance" as one half of a reciprocal exchange. No longer understood in the mode of *aid*, training was exchanged for access to research sites and restricted materials inside the pandemic epicenter (in particular, viral samples from birds).

As the report in *China HPAI Highlights* makes clear, the movement of data or expertise from one hand to another was not the only product of these exchanges. Much like guanxi gifts and favors, these exchanges also produced relationships—"working relationships," as the newsletter puts it. In turn, these relationships opened toward a future of subsequent exchanges, in which "similar field data collection efforts in the future are more likely to be accepted and successful." Rather than technical assistance as a form of development aid, technical transfer became a tool for building affinity and accessing the epicenter.

NETWORK ANALYSIS

When I first met Martin in 2010, he was in the midst of analyzing the results from a study he designed that examined the interconnections between live-poultry markets in southern China. At this time, many believed that live-poultry markets carried particularly high risk for avian influenza, primarily because of the crowded conditions at the market, the stress caused by travel that could lower immune resistance, and the mixing of species and sources of poultry. Many human cases of avian influenza had also been traced to markets.

Martin's project aimed to chart how markets were linked to one another and then, by using tools from social network analysis, examine whether the degree of connectivity was correlated with markets where H5N1 viruses had been isolated. In order to develop this analysis, the study needed two kinds of information: first, data on the trading connections between different live-poultry markets, and second, data on influenza outbreaks or the isolation of viruses in the selected markets. In the first preliminary analysis, Martin excitedly told me, a statistician found correlations between markets where H5N1 viruses were isolated and markets that were "k-cores" of the network—that is, hubs at the center of the network. In subsequent reanalysis, unfortunately, the correlation disappeared. Still, Martin was convinced that k-core was significant, that proximity to core (in this case, several poultry markets in Hunan Province) was more important than number of connections in the spread of H5N1.

As I spent more time with Martin at the Emergency Center, I quickly learned that networks—or more precisely, the cultivation of qualitative networks—played as large a role in the ethical and political process of data collection as they did in the scientific analysis of results.[46] Much as with the Poyang Lake study I mentioned above, Martin designed the live-poultry market study as an exchange of training for access to biological materials and information. The study took place in three provinces of southern China—Hunan, Yunnan, and Guangxi—that are included within the ecological region that is known as the pandemic epicenter. But there was a more pressing reason why these particular provinces were chosen: provincial Center for Animal Disease Control (CADC) leaders *agreed* to allow the training and research project to take place there, whereas wealthier provinces in the same ecological zone, such as Guangdong and Shanghai municipality, which already had well-developed veterinary inspection teams, refused offers of technical assistance.

Cai Haifeng led the data-collection portion of the study, along with the Emergency Center's national veterinary epidemiologist, Gao Lili. Cai and Gao traveled to each province, where they selected ten markets based on size, trade, and hygiene practices that were believed to be "representative" of the poultry marketing practices in those provinces. First, the Emergency Center staff trained both provincial and municipal (*shi*)-level CADC veterinarians in a standard protocol for collection of virological samples. During the following days, the Emergency Center staff and all of the trainees went to a local live-poultry market in order to demonstrate and exercise the sampling protocol. Typically, one CADC vet would hold a bird with its wings stretched back, in an effort to immobilize it, while a second CADC vet took a cloacal and tracheal swab. For the Emergency Center, the study design enabled access through aid: the training exercise involved the actual collection of samples from a live-poultry market for virological testing. Indeed, as Gao explained to me, because of China's decentralized bureaucratic authority, the training needed to include *both* provincial- *and* municipal-level CADC veterinarians: only the lower-level municipal bureau has formal authority over the markets within their jurisdiction, and thus including them was the prerequisite for access.

However, the negotiation of networks did not end there, for access on its own soon appeared to be insufficient. Although the training-based study design successfully accomplished the collection of cloacal and tracheal swabs for virological study, after collection the samples were moved to provincial CADC laboratories for investigation. Unfortunately, the provincial labs reported that all samples tested negative for influenza virus. (A similar result occurred in a Poyang Lake study conducted by Scott Newman on another occasion. As Newman noted about that other study, "I could believe there were no H5N1 HPAI viruses, perhaps, but no influenzas?") Skeptical of the result, Martin then negotiated for 30 percent of the samples to be forwarded to the National Key Laboratory for Avian Influenza in Harbin. Once there, Martin noted simply, "There were quite a few positives."

In addition to biological samples, the study also collected information on poultry trade patterns and hygiene practices in order to describe the "network" of live-poultry markets. Gao Lili led this effort, which focused on an interview-based survey of market managers and a select sample of poultry traders. A lengthy list of questions primarily aimed to track where traders sourced their birds and where they sold them, a heuristic for identifying connections among different markets. Although the survey did not involve a training component, Gao still brought two or three CADC vets along with her because she could not speak any of the local dialects. Gao was unsatisfied with the results of their translation, she told me. Perhaps because the CADC vets did not completely understand the purpose of her questioning or perhaps because they had their own motivations, by the time the answer came back to her, retranslated, it was compressed and almost useless. The vendors, to be sure, were not all that forthcoming. Probably, Gao felt, they were afraid that government authorities might close them down or were afraid about the theft of valuable business information. It did not help that the CADC officials accompanied her, with white lab coats, clipboards, and the rest. As a result, she had to develop her own interpretive strategies for validating the statements that vendors made.

When she asked the vendors how many birds had died, many would say that no birds died: "If people said zero, we knew it wasn't true. If they said five to eight percent, it meant there was normal death. If they said twenty to thirty percent, it meant there was an epidemic."

Gao started going to the markets at night, long after the official survey was over, and looking around on her own. One evening she encountered a transporter who sold poultry from his truck to the wholesale vendors at the market, who serendipitously happened to be from somewhere near Gao's own hometown in the northern province of Henan. Gao told me that she immediately began speaking with the transporter in a regional dialect, and as a result the conversation took on an intimate and free character: an intimacy not possible, Gao felt, in the Mandarin lingua franca. They were both far from home. And in the course of this conversation, the transporter told her something completely unexpected: he purchased his poultry in the far north of China, sometimes as far as Heilongjiang Province (more than 3,500 kilometers away), and sold them in the southern cities of Guangxi Province.

The encounter became almost legendary in the Emergency Center. It opened up a completely new understanding of the dynamics of poultry trade. It stretched the ecology of the so-called influenza epicenter far beyond southern China, where most previous research had located it. Gao's ability to translate from the dialect to the national and the international made the otherwise invisible extent of unofficial poultry trade visible. The encounter produced a distinctive kind of truth statement, one that constantly circulated inside the Emergency Center, guiding the interpretive meaning ascribed to the analytics of poultry trade networks. In public presentations, Martin often spoke of it too. In one public talk, for instance, when Martin presented the results of the network study, he explained that even before they conducted the quantitative analytics, the team "could see already that there was some long-distance trade going on in several provinces." Yet Gao's finding could never become a true scientific fact, in the sense that it would never be included in any published articles.

In the previous section I described the boundary between international and national inside the Emergency Center. In the live-poultry market study this boundary played an important role in the center's division of scientific labor. The more senior national staff members, Cai Haifeng and Gao Lili, collected the data "in collaboration," as they called it, with provincial and municipal veterinary services. The Chinese Harbin Key Laboratory for Avian Influenza sequenced the virological specimens and sent digested reports to the Emergency Center. Once the sequences and accompanying surveys had been delivered to the Beijing Emergency Center, however, Martin conducted all of the analysis, with the help of Ricardo Magalhães, an "international veterinary epidemiologist based in Queensland University."

Magalhães came to China for only brief visits of five days or less. He explained to me that he plans these trips carefully in order to accomplish all of the tasks that will require working with national staff during the short stay. Most of the numerical and statistical analysis he works on by himself back in Australia. When I met him on his last day in Beijing, he worked until six "wrapping things up"; then we all went to the famous Silk Market in Sanlitun, where a national staff member helped Magalhães buy toys and other inexpensive souvenir products for his family before dinner. The lower levels of prestige awarded to technical data production, as opposed to analysis and journal-article writing, lies deep in the modern sciences.[47] In the projects of the Emergency Center, this hierarchy of prestige is mapped onto the distinction between national and international status.[48] When Gao Lili explained the results of the study to me, she reaffirmed that a deep separation divided the production of data from the methodology of analysis. She described her experiences at the markets in some detail, explaining how she collected virological samples and conducted surveys of poultry traders. She confidently explained the results of the study to me. She then noted, however, that she "won't speculate or comment on the statistical methodology employed by Martin and Ricardo." Moreover, although both Cai and Gao are included as coauthors in the resulting publications, the other "collaborators" from national, provincial, or municipal veterinary institutes are not.

Gao Lili's own career, however, reveals that this boundary between international and national is not a fixed ascription of status, but rather sets people and things into motion. Gao was born in Henan and took a bachelor's degree at Henan Agricultural University, graduating in 2000. After graduation, she began working as a lecturer at the Henan Institute of Science and Technology. She soon enrolled in a master's program in preventive veterinary medicine, studying with Chen Hualan at the National Key Laboratory for Avian Influenza in Harbin. In 2005, following the reemergence of the H5N1 virus, the Ministry of Agriculture sent her abroad to study for a master's in veterinary public health, a degree at that time not offered in China. She attended a joint program held in part at the Free University of Berlin and in part at Changmai University in Thailand.

When I worked with her at conferences or training programs, Gao more than once remarked on the different pedagogical style she had discovered abroad. In China, veterinary science is taught almost entirely through textbooks. This was a mode very familiar to her, both from her time as a student and through her own teaching experience. But in Berlin, she recalled that the professors assigned students to conduct field investigations. In one exercise, for example, they asked students to go to slaughterhouses to learn about food-safety regulations and how they were administered in practice. In Germany each class had fewer students than is typical in China. Classes were structured in a seminar style, with students and the professor seated at a round table together discussing the topic of the day. Finally, rather than lecturing on facts and theories for the students to commit to memory by rote, the professor often began with a problem and asked the students to come up with a solution.

Gao took this "international" standpoint in order to criticize China's veterinary education, and she sought access to the international domain—through special training programs and the like—in order to increase her own abilities as a veterinary scientist. At the same time, in projects such as the live-poultry market study, she also cultivated a distinctive position at the boundary of the national and the international that enabled her to produce unique scientific values. In Gao's own career, the Emergency Center therefore provided two mechanisms of value creation. On the one hand, the international stood as a source of value for Chinese veterinarians like herself; by going to Germany to study and even by working at the Emergency Center, Gao could increase her own value as a veterinary scientist. On the other hand, the Emergency Center depended on her to remain on the *national* side of the boundary line, because it was from there that she could best translate local situations into globally circulating truths. As a result, these two mechanisms for mobilizing the boundary between international and national sometimes stood at odds. Having accumulated enough value through her work of translation, Gao suddenly left the Emergency Center in the late summer of 2011, while Martin was away on vacation. She had been recruited by another international veterinary expert (who had recently visited the Emergency Center on an FAO mission) to study for a PhD in veterinary epidemiology at a Canadian university. When Martin returned to Beijing and found she had left, he was shocked and disappointed. In his eyes, he explained to me, Gao had revealed herself to be "opportunistic" by moving onward as soon as a better situation appeared. But it was the Emergency Center that had created the opportunities.

FIELD GUIDE TO THE EPICENTER

Some time after he established the Emergency Center in Beijing, Vincent Martin, always mobile, began studying for a PhD in spatial ecology under the supervision of Marius Gilbert at the University of Brussels Spatial Epidemiology

Lab. As part of his dissertation research, Martin collaborated with Gilbert to develop a spatial risk model for China, comparable to the one Gilbert had developed for Thailand. However, the invisibility of silent virus circulation made the development of this spatial ecological risk model quite different from Gilbert's earlier models.

First, the data necessary for building the models were much more difficult to access. As I describe in chapter 2, Gilbert's Thailand study was based on data collected by the incredibly thorough active-surveillance program known as the X-Ray Survey. The data were collected over a one-month period during the second-wave epidemic in 2005. Active surveillance means that the teams actively sought out cases in every district rather than waiting for reports of outbreaks to trickle in. Moreover, the X-Ray Survey combined laboratory-supported surveillance for H5N1 avian influenza with a simultaneous survey of poultry populations, another crucial data source. Neither virus surveillance nor poultry surveys were as comprehensive, consistent, or accessible in China, particularly in the more recent years when no outbreaks were reported. Poultry population statistics, for example, vary from province to province in terms of the how animals are categorized, with some provinces reporting numbers that distinguish chickens and ducks, and others reporting only aggregated poultry numbers. As a result, in the China study Gilbert relied on numbers from the FAO's global livestock population modeling program, "Gridded Livestock of the World," rather than provincial census data.

On the other hand, the silent circulation of the H5N1 virus also meant that the purpose and utility of making an ecological risk map was quite different. In Thailand, widespread culling of poultry discovered during the X-Ray eliminated the H5N1 virus. It did not return. By the time Gilbert began his Thailand study, he was working with archival data that represented what would best be described as an archival situation. As a result, Gilbert's study was not designed to provide tools for epidemic control in Thailand. Rather, Gilbert aimed primarily to produce what he called "generalisable" tools and facts about avian influenza risk.[49] Through the authority attributed to his models by scientific publication, Gilbert hoped to construct scientific facts that would be true not only for Thailand but also elsewhere. As Gilbert and colleagues state, "The model of HPAI H5N1 virus risk developed in Thailand with the data from the second epidemic wave maintains its predictive power when applied to other epidemic waves or other regions, indicating that the model can be extrapolated in space and time."[50] To this end, Gilbert "validated" the study by using the model to develop and test risk maps of Cambodia, Laos, and Vietnam.

In the extrapolation of the model to China, although facts could still be constructed and articles published, the primary purpose and use of the resulting maps were quite different. Whereas HPAI H5N1 was only a historical, archival problem in Thailand, in China avian flu was becoming more or less an endemic problem. At the same time, the circulation of viruses was largely invisible and unreported. Given this situation, the Emergency Center staff used the maps of risk distribution *as a heuristic substitute* for outbreak reports in order to guide the location of biosecurity interventions into the pandemic epicenter. In a characteristic shift, the modeling of ecological risks substituted for the bureaucratic chain of surveillance and outbreak reporting.

When Martin invited and supported Gilbert's geo-spatial risk model in China, he was not only interested in building models that could lead to general facts about influenza risk (for instance, demonstrating the hypothetical importance of free-grazing ducks). Rather, Martin hoped to locate zones of high risk inside China. Along with Gilbert, he hoped to create an analytic device that could orient where to conduct scientific research and guide practical interventions on the hypothetical source of pandemic influenza. This was not so much a model of the pandemic epicenter as field guide.

The China geo-spatial risk model took risk factors previously identified as significant in the Thailand studies—for example, duck populations and rice-cropping intensity—and tested their correlation with archival data on influenza distribution in China. These data come from the *Veterinary Bulletin*, a newsletter published monthly—in Chinese—by the Ministry of Agriculture. The data include two types: reported outbreaks of HPAI H5N1 for the years 2007 to 2009, and the positive samples of H5N1 collected by the Harbin lab during postvaccination surveillance over the same period. Because it is published only in Chinese, its contents remain limited to Chinese readers. In the case of the risk map study, therefore, a key—but low-prestige—data-analysis role was provided to the "national" staff members of the Emergency Center, especially Cai Haifeng. Their task was simply to translate the data points from the Chinese-language charts in the *Veterinary Bulletin* into an English-language spreadsheet. In the final paper, Cai is credited as having "contributed reagents/materials/ analysis tools."[51]

Once Marius Gilbert had entered these data into the models, the results revealed a far more heterogeneous terrain than previously imagined in the hypothesis that southern China is an influenza epicenter. In his original hypothesis, Shortridge had typically emphasized the Pearl River Delta region adjacent to Hong Kong, with its "intricate waterways," as the core of the influenza epicenter. Based on the risk model, Martin and his colleagues concluded that the epicenter could actually be divided into three distinct zones, each with a different level of risk. They write that their new model "supports the hypothesis of a wider and slightly displaced epicenter of influenza viruses, not only concentrated around the Pearl River Delta in Guangdong province but extending south of the Yangtze River and including provinces such as Jiangxi province where internal segments of the 1996 geese HPAI H5N1 virus may have originated."[52] Indeed, the distributed clusters of red on the map, each indicating a zone of elevated risk, suggest a multiplicity of localized epicenters.

But it is the use of the map that I want to focus on here. In early 2011 I helped the Emergency Center organize a bilateral meeting between the agriculture ministries of China and Vietnam on the topic of "targeted risk-based vaccination." During the course of this conference, I saw Martin deploy the map in a public setting for the first time. In the process, I saw another strategy for accessing the epicenter, one that supplanted political, territorial space with representations of ecological space.

In typical style, Martin planned the "China Vietnam Forum on HPAI Management and Control" as a mechanism of persuasion. This was no gathering of free scientific exchange, no symposium to debate various views of a problem. Martin organized the entire meeting to make an argument: that China should replace compulsory universal vaccination with "targeted, risk-based vaccination." The method, previously demonstrated during the rinderpest eradication campaign in Africa, aims to identify areas of high risk in order to focus scarce vaccination resources. Rather than vaccinating everywhere in policy (but likely unevenly in practice), the targeted approach is intended to ensure comprehensive vaccination over a more limited set of high-risk areas.

The forum took place at a large international hotel in Sanlitun. An enormous ballroom with a stage was outfitted with round tables, around which clustered different groups of participants—international technical experts at one table, FAO international staff at another; visiting representatives from Vietnam at one table, and representatives from the different agencies of China's Ministry of Agriculture at several other tables, including the Veterinary Bureau, the Center for Animal Disease Control (CADC), the China Animal Health and Epidemiology Center (CAHEC), and the Harbin Avian Influenza reference laboratory. At the height of the meeting an unexpected controversy erupted over the efficacy of the universal vaccination program. At the time the controversy surprised me because it did not take the classical form often described in accounts of pandemic flu research, in which international criticism confronts China's assertions of national sovereignty. Rather, this time the controversy divided two parts of China's agricultural bureaucracy against each other: the Ministry of Agriculture's CADC, represented by Wang Hongwei, and the Harbin lab, led by Chen Hualan.

In his presentation on China's poultry immunization, Wang Hongwei emphasized the success of the vaccination program. He explained that each year the Ministry of Agriculture issued a national HPAI surveillance plan. The plan included postvaccination serological surveillance, in which poultry are sampled not for viruses but for the presence of antibodies produced by vaccination. The provinces organized the implementation of this plan, and laboratories at the province and prefectural level—along with the so-called national monitoring station network (*guojia cebao zhan*)—carried out the surveillance. The results of every sample were entered into a central computerized system run by the CADC called the "Nationwide Animal Disease Surveillance and Epidemic Information System" for integrated analysis. In 2010, for example, the surveillance program collected more than 4.3 million serological samples and 347,000 viral samples. In conclusion, Wang happily reported that the CADC's serological surveillance showed that China had achieved herd immunity—defined as 70 percent positive rate or better—in every province. As the concept of herd immunity indicates, Wang claimed that China had vaccinated enough poultry to effectively break the circulation of the virus and prevent outbreaks from occurring.

Later that same day, Director Chen Hualan of the Harbin Key Laboratory presented a paper discussing the work of her lab in the development of vaccines—a paper that offered a very different assessment of the vaccination apparatus. Postvaccination surveillance, she explained, is an important part of their vaccine development program. But its purpose differs markedly from the official national surveillance program administered by the CADC. Chen explained that the Harbin lab conducts surveillance not to assess the coverage of the immunization campaign across national space but rather to detect vaccine failure—that is, to discover circulating strains that have become antigenically diverse to the current vaccine formula, rendering the vaccine ineffective. When discovered, such circulating strains typically become the seed virus for new vaccine development.

Chen stated flatly that, based on postvaccination surveillance, "vaccination coverage is definitely not as high as 90 per cent. In broilers over 80 per cent were not vaccinated. In ducks, over 70 per cent were not vaccinated according to our surveillance." Whereas Wang suggested that at least 70 percent of poultry were vaccinated, Chen seemingly inverted the claim, arguing that over 70 percent were *not* vaccinated.

As the dispute unfolded over the course of the day, it became clear that the discrepancy was rooted in the different purpose and design of the surveillance programs. The two surveillance programs are guided by quite different means and ends. The CADC surveillance uses sampling devices to provide a representative assessment of the mass vaccination program across the entire territory of China. Relying on the hierarchy of the veterinary bureaucracy, the ACDC sends county-level veterinary offices to conduct or collect the results of serological tests from farms, assembling a systematic and regular distribution of samples across space. The Harbin lab, by contrast, sets out to find the *gaps* in the vaccination program. Rather than assessing the coverage of the immunization campaigns, the Harbin lab aims to detect vaccine failure. As a result, it samples for viruses at live-bird markets, not farms. When such antigenic variants are discovered, the Harbin lab updates the vaccine seed strain to include the new antigens and produces new vaccines with the updated seed strain. "New strains" are discovered precisely when a virus is found to escape the coverage of existing vaccines; novelty, in this case, is merely the externality of the vaccine's efficacy. And alongside the progressive differentiation of viral variation, the Harbin lab has produced a similar progression of vaccines: as of 2011, Re-1 had been joined seven other vaccines, culminating with the production of re-8.

A key exchange took place when the floor was opened for questions. Chen acknowledged that "when we went to farmers to collect samples, we see the same result [as the CADC]. When we go to markets, we see different results. For me it's very easy to explain. The local people in charge of vaccination, they know which farmers are vaccinating properly, so they bring us to those farmers; it's also where they are collecting samples."

Chen suggested that the base-level state-employed veterinary workers selectively pick "good" farms—farms they know are "vaccinating properly"—for serological sampling, thereby skewing the results. From the audience an official from the China CADC offered a different interpretation, one that focused on the challenge of veterinary work amid China's complex diversity of husbandry systems. "Backyard and grazing[53] farms, they maybe only get one shot," he said:

> If you have been to Chinese markets, you will see dozens of chickens, not a lot, never over a hundred. Our surveillance rates are also lower than that reported by local government. Agriculture ministry is also concerned about this. [There are] different models of agriculture in China. How do you sample? If you sample commercial farms, you might have a high rate, but if you sample backyard farmers, the rate will be lower. In the policies,

before the birds go to market, there is no requirement to [re-]vaccinate. The over 70 per cent titer is because these numbers are monitored at the farms. The birds sampled by Dr. Chen [at markets] probably hadn't been vaccinated for over four months.

Amid this conflict over the comprehensiveness and efficacy of the universal vaccination program, Martin took a detour. Rather than vigorously asserting the failure of the vaccination program or the limitations of surveillance mechanisms, he turned to the risk map as a way to bypass controversies over reporting and vaccination. In his own presentation Martin made the case for a transition toward what he called "risk-based targeted vaccination." He reminded the audience that the goal of the H5N1 response—stated clearly in a formative OIE/FAO strategy document—was the *eradication* of the highly pathogenic virus. Eradication meant total elimination of the virus, not merely control or reduction of outbreaks. And evidence indicated that mass vaccination could never eradicate the virus because of persistent silent circulation.

"China is huge," Martin told the assembled crowd, "and we still don't really know where are the main risk zones." With an image of the China risk map projected on a large screen hanging from the ceiling, filling the wall behind him, he continued: "The red areas are high risk, and this shows it is not all of south China, but that it is associated with particular risk predictors. . . . Thanks to this analysis you can concentrate efforts, you can use this map to overlay outbreaks, human cases. You can see that lots of parts of China have very little risk."

In Martin's hands the risk map could be used to shift vaccination efforts from the universal coverage programs designated by state policy, which arguably could never be achieved in practice, toward distributed terrains of relative influenza risk. Perhaps more importantly, the risk map also allowed the Emergency Center to bypass the controversial questions about vaccination coverage, the silent circulation of viruses, and the lack of transparent outbreak reports in its planning of research and biosecurity interventions. Answers to those questions, or more precisely, the forms in which an answer to those questions could take, were inevitably tied up with the politics of national sovereignty and the hierarchy of administrative bureaucracy. The risk map, by contrast, allowed the Emergency Center to focus activities on certain areas and ignore others without relying on outbreak reports or effective surveillance systems. Indeed, risk maps of the spatial ecology of the influenza epicenter could now be made with data collected from a satellite flying seven thousand kilometers overhead.

Unlike in humanitarian renderings of global health, Martin did not figure a planetary humanity as an ethical imperative for bypassing sovereign states but rather aimed to produce a technical demonstration of the insignificance of national scale or territorial boundary.[54] Put another way, the rooting of risk maps in satellite-derived data invoked a "vertical" geopolitics explicitly designed to supersede national territorial boundaries.[55] One could say he attempted a technique of stratification that would separate China—understood as a sovereign, territorial entity in geopolitical space—from the ecological layers of the pandemic epicenter.

PREDICAMENTS OF CULTURE

Even as he upheld the technical over the political, the risk map drove Martin and his collaborators back into the epicenter and its politics of affinity. For one thing, the validation of the remotely sensed images produced by orbiting satellites constantly required on-the-ground surveys, usually conducted by local collaborators. In remote sensing this process is called "ground-truthing" and involves calibrating the grid of pixels in a satellite image with on-location observations tagged to other rectangular grid systems, such as longitude and latitude.[56] In 2011, for example, a research team led by the China Academy of Sciences in conjunction with Jiangxi Normal University traveled across the Poyang Lake region. The team mapped the boundaries of fields, using hand-held GPS devices, and interviewed farmers, primarily asking them whether they planted one, two, or three crops of rice per year. With these surveys the research team aimed to demonstrate that the remote sensing "signature" they had proposed, based on the temporal dynamics of the normalized difference vegetation index (NDVI), could accurately distinguish single from double cropping of rice in Landsat satellite data.[57] The vertical perspective orbited overhead, but its interpretation relied on building affinities with local collaborators, crossing marshy ground in white vans, and exchanging knowledge with farmers inside the epicenter.

Moreover, Martin's performance of *stratifying* ecology from politics relied on the strategy of affinity as its foundation, its infrastructural support. When I joined several of the invited speakers and international FAO consultants at a Latin-themed bar on a trendy Beijing *hutong* street after the China Vietnam Forum, everyone agreed the meeting was a remarkable success. Two of the FAO consultants from the Vietnam Emergency Center in Hanoi even remarked that it was one of the most successful meetings they had ever attended. Knowing how Martin had clear objectives for the meeting, I remarked that the whole thing must have been orchestrated very effectively to achieve them: almost everything Martin had hoped for had ended up written in the official "recommendations" issued after the forum, and granted ministerial-level approval by both China and Vietnam. Martin then turned to me with a smile and said that he wrote the recommendations *before* the meeting:

> I had started thinking about what the recommendations should be when S. Y. [a Ministry of Agriculture leader] called me on the phone and said "I think you should write some recommendations." I answered, as if surprised, "Really, you want me to write them now?" and S. Y. was a bit shy, you know, but we agreed to do it. Then I had to work late Monday night to make sure I could get it done, and send it around to S. Y. to look over before the meeting. Then as soon as the meeting started I sent them around to the Ministry of Agriculture, and they made a few changes.

The FAO consultant from Vietnam chimed in: "Yes, we always do this, there is some kind of moral thing about waiting until the meeting has officially begun, but once the meeting begins you just send in the recommendations. The Vietnamese and Chinese like it this way; they expect it."

Strategies of affinity, such as the banquet dinners hosted or attended by the staff of the Emergency Center or attempts to do things the "Chinese way," did more than break down barriers to transnational communion. It is more accurate to say that they marked borders in a way that enabled communication across difference. This drawing of borderlines does work of its own, providing a language through which difference can be understood and worked with. And for Martin, what he described as Chinese "culture" was an ambivalent but central part of his work, attractive and energizing but also troubling and exhausting.

For most of his time working at the Emergency Center, Martin lived in one of the oldest sections of the city, in a hutong near the Temple of Confucius and the Yonghe Lama Buddhist Temple. Along with his wife and children, he lived in what was called a "villa," a modern construction nestled inside a maze of alleys and courtyard homes. Later he moved to a traditional courtyard house near the North Lake, in the very heart of the old city. An ancient structure, the inside had been painted blinding white and the wooden beams varnished by the English real estate investor who rented it to him. Recently installed glass walls closed off the rooms from the inner courtyard. The furniture continued the theme of *chinoiserie* but was juxtaposed with a collection of African sculptures Martin had collected while working on the eradication of rinderpest in Kenya. One evening in early December 2011 we sat together enjoying French wine and cheese in the main room. My fieldwork was coming to an end, and Martin seemed tired. "These hutong buildings are impossible to heat," he muttered and pulled a space heater closer to us. He told me he was planning to leave China for good in June, to take up a position in the FAO headquarters in Rome. His family had already returned to Europe.

When he reflected on his four years in China, Martin told me, he found a "contradiction" ran through all of his work with Chinese counterparts, including in his own office. He recollected projects like the Poyang Lake virus sampling expedition, in which baijiu parties led to unprecedented access to samples in a key avian influenza site. "In a place like Cambodia and Laos, even though at first you might have more open access, you could end up hitting barriers, not be able to get as far as here," he explained. But in other moments he was frustrated that China still did not publicly report certain outbreaks or rejected projects that he proposed. When Martin returned from Thailand after the New Year, the moment that I began this chapter with, he found a regional foot-and-mouth disease (FMD) crisis that China claimed did not exist within its borders. Outbreaks had been reported in Mongolia, southeastern Russia, and even in the highly secretive North Korea. But China had not reported a single case. Despite the fact that the highly contagious virus had been found along all of the international borders around China's northeastern province of Heilongjiang, there was no indication that any cases existed within China's territory. From a regional and ecological point of view, Martin felt, the virus *must* be in China as well. "I think we are going backwards here," he lamented at the time.

Martin saw "culture" as a force that shaped this contradiction. In one email exchange, I had pointed out how my research as an anthropologist explored science as a cultural practice. In his response, doubtless conscious he was speaking to an anthropologist, Martin picked up the terminology of culture to identify Chinese "cultural practices" as a blockage to full and free communication:

I agree with you regarding the cultural practices, and there was another dimension I was thinking about lately, which is the way countries and government deals with sanitary information and their own cultural approach to what we call "transparency." I am not sure it would lead to any sort of interesting findings, but I am really thinking sometimes (especially when we talk about China) that this notion of transparency and information sharing might need to be revisited in light of cultural practices (or maybe it has already been done). In the same vein, the cultural and ethical approach to science is also of interest: we know that hiding a disease or not sharing virus sequences is easier (in the short term) than doing it, and I was also wondering whether we are missing a point, as Western countries, when we pretend that everybody should play the game and be transparent.

Martin employed the notion of cultural difference in order to explain China's seemingly contradictory approach to working with global health agencies. The concept of culture enabled him to constitute China as different but fully *reasonable*. Moreover, he advanced a kind of cultural critique, suggesting that China's difference obligates us to "revisit" our own notions of transparency and information sharing. Francois Jullien has remarked on the "indirect" approach to meaning in Chinese discourse, which he claims allows us to reflect on the boundaries within which Western thought has grown. In particular, Jullien argues that Chinese "detours" in discourse reveal a mode of thinking that does not prioritize language as a representation, more or less transparent, of nature or being. To be clear, I do not draw on Jullien's *Detour and Access* in order to bolster observations of Chinese culture with philosophical antecedents. Rather, I aim to elucidate how Martin himself, in a rhetorical move that resembles Jullien's more than a little, attempted to make the difference of Chinese culture "intelligible." With the terms of cultural difference, the contradiction Martin faced in China no longer was only a blockage but also could be made productive. Understanding the Other in terms of cultural difference opened an entire register of practical work. In fact, by building relationships upon the basis of cultural difference, things could be communicated that couldn't be said.[58]

After the FMD crisis of March 2011, the EU wanted to develop a training program for FMD, but China's Ministry of Agriculture rejected the proposal without explanation. At the request of the EU's diplomatic representative in Beijing, a Frenchwoman who strongly supported the work of the Emergency Center, Martin got involved. As he was preparing the materials for the course, he told me:

> I found that you cannot say "FMD" here. When you talk about FMD, people just say, what are you talking about? But there are some people I know at MOA that I can talk to sometimes. It is not like there is simply no possibility of people talking; in certain situations they might speak very frankly. *You have to talk like it's a joke or whatever.* And so I talked to these people I know and said, what is it about FMD, what is the problem with FMD? And they said there was no problem. But the problem it turns out was with the training program, because they said: we don't want to have experts fly in and say things to us we already know, or are too basic. It's true, the Chinese are too polite; they would nod their heads and smile and at the end of it say

"Thank you very much," but at the same time in the back would be saying how ridiculous this was. In fact, I agree; I once saw an EU training where I was like, guys, why are you talking at this general level? There are concrete things that [the] China MOA side want[s]. . . . This is really my role. In other countries, EU will offer a lot of money for technical assistance for FMD without a lot of specifics attached, and people will take it. But here it will be refused. So I first go to the China MOA and find out, what is it that you can say, what is it that you want? Then I go to the EU and try to get them to say and offer exactly that.

Martin's everyday diplomacy was not limited to the apparent opposition between international objectives and national state interests, or the territorial struggles among global health norms and sovereign powers. Rather, by following him as he established the Emergency Center in China and moved into the epicenter, I saw how he worked in, around, together with, and beyond the geopolitical strata of national biosovereignty: including China's reluctance or inability to report virus surveillance, the efficacies and inefficacies of a universal poultry vaccination program, and the institutions and norms of veterinary practice. At different moments Martin employed distinct and even seemingly opposed strategies to access the epicenter: the cultivation of networks of affinity with Chinese ministry officials, technical experts, and veterinarians, on the one hand; and the split and separation of ecological truth from political interest, on the other. At the beginning of the chapter I noted that Martin's appraisal of China's willingness to cooperate veered between optimism and outright denunciation. Martin's judgments and critiques are, I now suggest, conditioned as much by his choice of access strategy (affinity or stratification) as they are by the attitudes of his counterparts in China (cooperative or antagonistic). I would also suggest that despite Martin's occasional and situational mobilizations of denunciation, the movement into the epicenter relied on affinity as the ultimate ground for negotiating the geopolitics of national biosovereignty. Now, in the last chapter, I will turn to discuss Martin's most ambitious effort to cultivate access through affinity: the FAO Emergency Center's high-level training programs for China's state-employed veterinarians.

CHAPTER SIX
OFFICE VETS AND DUCK DOCTORS

"To understand China's veterinarians, you need to know political economy," Cai Haifeng, the senior national veterinarian at the FAO Emergency Center, told me. We talked over lunch at an upscale Yunnan-style restaurant filled with white-collar professionals and expats near the Emergency Center offices. Before joining the FAO, Cai had spent decades working as a livestock veterinarian battling epizootics, primarily with the Animal Quarantine Service (recently renamed the China Animal Health and Epidemiology Center). He was among the first generation to study veterinary science in the post-Mao period. Although his current mode of life found him more often in urbane settings such as this restaurant in Sanlitun, his commentary drew on his direct experience with treating the pathologies of China's rural transformations.

Since the market reforms, Cai explained, the production of animals had grown rapidly, but veterinary expertise had not kept pace with the livestock revolution: "The relations of production and means of production are not being reflected at the level of ideology." Cai's account differs from those that identify the expansion of livestock production as pathological in itself, such as those that blame industrial broiler operations for the emergence of novel avian influenza viruses.[1] He points to the dynamic historical relations between knowledge and practice, between the science of animal disease and the means of livestock production, and locates pathology in their lack of correspondence.

I heard similar diagnoses from several other veterinarians, including Ning, another national staff member of the Emergency Center. China's livestock disease crisis emerged because the "quality" (*suzhi*) of veterinarians was too low, Ning told me one day at the Center. The veterinary "software" was no longer adequate to China's rapid material growth in livestock production, its developmental hardware. To extend this metaphor of code overwhelmed by matter, he compared the veterinary condition with the notorious "ghost cities" of China's real estate boom. Bricks and mortar can be laid very quickly, he said, but too often the buildings are left empty; no one lives inside. In much the same way, livestock production has grown rapidly but lacks qualified experts to care for the animals and their diseases.

This chapter examines the FAO Emergency Center's efforts to improve the abilities of China's state-employed veterinarians through the Field Epidemiology Training Program for Veterinarians (China FETPV—Chinese translation *zhongguo shouyi xianchang liuxingbingxue peixun*). Upon my first arrival at the Emergency Center, Vincent Martin had told me that the current focus of his work was developing this two-year course in field epidemiology. "Field" epidemiology aims to cultivate "first responders" for epidemics, providing training in epidemiological tools for immediate outbreak investigation and response rather than academic research. As Martin explained to me, "There were plenty of veterinarians [in China], a lot were trained in laboratory technique, some in epidemiology, but epidemiology was really not the topic that was strong in their curriculum." The FAO first created an FETPV program in Thailand in 2008, adapting the content and form of the course from the U.S. CDC's Field Epidemiology Training Program (FETP) for workers in the human public health sector, which was itself the international export version of the CDC's famous Epidemic Intelligence Service (EIS).[2] China's FETPV was to be the first veterinary program of its kind organized at a national, rather than regional, scale. In a presentation to the Ministry of Agriculture in January 2010, Martin and Regional FAO epidemiologist David Castellan had outlined the "needs, vision, and goals" for field epidemiology training in China. As they pointed out, "China faces complex geographic and demographic challenges in controlling animal diseases. For this reason, capacity building must be based on a proven training model, provide minimum standards and also be flexible and adaptable. China needs a comprehensive FETPV to bridge the gaps in veterinary epidemiology training in this country." At our first meeting, Martin had been blunter: "Chinese vets are very bad at epidemiology."

In the emerging global health regime of pandemic preparedness, China has been marked out as a double locus of planetary danger: not only a hot spot of emerging viruses but also one of the "weak links in global preparedness."[3] In the previous chapter I focused on how the FAO Emergency Center aimed to transform the political obstacles to outbreak reporting and data sharing through the cultivation of new forms of cross-border collaboration. But as public policy scholar Joan Kaufman argues, China not only "lacks transparency in acknowledging outbreaks," but also, perhaps more alarmingly, "its health care system is not up to the task of putting in place systems to ensure preparedness of the capability to contain the epidemic if it begins in rural areas." A particular "area of concern," Kauffman notes, "is China's veterinary service and its ability to control" outbreaks of disease.[4] In other words, the danger that global health agencies attribute to the pandemic epicenter—why this particular site is considered a threat—is rooted not only in the pathogenic landscapes that drive the emergence of disease but also in the basic technical capacity of veterinary services to detect and contain a disease that emerges. The resolution of this double locus—and indeed the long-term effort to remediate the pandemic epicenter—has come to center on the ethical and technical improvement of China's veterinarians.

PROFESSIONAL STRATIFICATION

The first day of the inaugural FETPV module took place on a frigid Beijing December day in 2010, the kind of day where the city smog blocks out the sun and a midnight chill stretches far into the morning. I had trouble finding the offices of the Beijing Animal Hygiene Inspection Service (*dongwu weisheng jiandusuo*), located in an obscure corner of Beijing north of the zoo and practically underneath an overbuilt flyover, and was running late. I hurried up the unheated stairs to the top-floor conference room, passing on my right a large poster displaying photos of each of the Inspection Service officers dressed in crisp, dark-blue uniforms. When I opened the doors, I found the sleekly designed conference room, decorated in muted grays and blacks, already packed and very warm. The fifteen state-employed veterinarians selected as trainees sat around modular curved tables outfitted with built-in microphones, each staring into brand-new black Lenovo laptops gifted as both perks and needed tools for their participation in the training course. Most of them worked at Centers for Animal Disease Control (CADC) or hygiene inspection offices at the province level, although several came from national veterinary research institutes such as the Harbin Key Laboratory for Avian Influenza and the China Animal Health and Epidemiology Center (CAHEC).

At the front of the room, Vincent Martin was just finishing his brief opening remarks in English, with a translator nearby. Several European men and women who had flown in to Beijing to teach the training course stood beside him, including Dirk Pfieffer, a senior professor of veterinary epidemiology and the lead trainer for the first module. The national employees of the FAO Emergency Center sat on folding chairs around the edges of the room, alongside several "mentors" from Chinese veterinary institutions such as CAHEC, as well as other Chinese and international observers. I found a seat next to a rail-thin American who hardly spoke, whom I later learned was the Asia representative from the primary donor, the U.S. Agency for International Development (USAID).

After Martin finished, Pfieffer began to provide a conceptual overview of the course. Pfeiffer claimed that across the world, veterinary medicine faced an epochal moment. When we think of the vet, he said, we usually imagine the kind man or woman in rubber boots out at the farm, inspecting a sick cow or horse. The vet is traditionally concerned with the health of the individual animal. However, the recent increase in emerging zoonotic diseases—from mad cow to bird flu—has put this doctor of individual animals at a loss. Today, Pfeiffer suggested, animal disease must be addressed at the population level, the only scale at which emerging outbreaks can be identified, prevented, and eradicated.[5]

As I soon found out in conversations during coffee and lunch breaks, the Chinese veterinary trainees enthusiastically accepted Pfeiffer's claims that veterinary medicine was in crisis, faced by unprecedented disease threats, and that epidemiology was the answer that they needed. To them, however, this was not really an epistemological challenge based on the difference between individual and population approaches to disease. After all, the trainees came from state veterinary departments that had long been focused on disease eradication and control of epizootics. Rather, they argued that the specific history of modern China's development and its post-Mao market reforms created a distinctive crisis for veterinary governance and the character of the veterinarian. As a result, they understood the FETPV program not only as an instance of technical transfer from an international agency but also as an exemplary component of a broader process of national veterinary reform.

Although the veterinary trainees all came from elite institutions, at the province or national level, they told me that their profession as a whole compared poorly with the status held by other sciences in contemporary China and, in particular, human public health. "Very few veterinarians get to study epidemiology," one trainee explained to me. "We're still very behind." This had not always been the case, as many older vets recalled: during the 1950s the new Communist government held up veterinary work as a crucial component of rural socialist construction. In those days the government worked to unite and train existing animal health specialists, referred to as "folk vets" (*minjian shouyi*), within institutions of state government such as Animal Husbandry and Veterinary Stations (*xumushouyizhan*, AHVSs). By 1959, the government had established an AHVS in every Commune; by 1979, the state employed as many as 700,000 veterinary experts and anti-epizootic personnel at the township and village level. In the 1990s, however, the AHVS system fell into disarray, with many township and village vets seeking income from the private sector or abandoning veterinary work altogether.

In the early 2000s, in part because of requirements impinging on China following entrance into the World Trade Organization (WTO), leading Chinese veterinarians began calling for reforms to the veterinary service. These governmental reforms ranged from establishing a national-level veterinary bureau in the Ministry of Agriculture and appointing a chief veterinary officer to the wholesale structural reform of the veterinary bureaucracy. At its core, the veterinary reform aimed to distinguish the massive number of state-employed veterinary workers into two groups, selecting a smaller number to retrain and recertify as a new kind of "official veterinarian" (*guanfang shouyi*) while leaving the rest to find their way in private practice. Although already partly underway by 2002, the SARS and avian influenza outbreaks exposed the failure of the existing veterinary system and demonstrated the urgency of veterinary reform.

Katherine Mason has described a similar process of reform in China's human public health sector.[6] Beginning in the late 1990s, the U.S. Centers for Disease Control began training China's public health workers, particularly in the area of epidemiology, while China embarked on a large-scale reform of its disease-control infrastructure based on a CDC model. Although this process was already underway before SARS broke out, the epidemic played a crucial role in shaping and encouraging the process of public health reform. These similarities between human and animal health reforms are not coincidental. The vets I trained alongside of held up the public health sector as an ideal model of reform, and many aspects of the veterinary reform are explicitly based upon the reforms in the public health sector. For example, just as in the public health sector Centers for Disease Control and Prevention, or CDCs (*jibing yufang kongzhi zhongxin*), replaced Mao-era Anti-Epidemic Stations (*fangyizhan*), so in the veterinary sector Centers for Animal Disease Control (*dongwu yibing kongzhi zhongxin*)[7] replaced Animal Husbandry and Veterinary Stations (*xumu shouyizhan*). Broadly speaking, this transition reflected a shift from a Soviet to an

American model of health governance. In another specific mimesis, the Emergency Center's Field Epidemiology Training Program for Veterinarians (FETP-V) is adapted, as I noted above, from the U.S. CDC's international outreach Field Epidemiology Training Program (FETP), including the Chinese incarnation of the FETP that played a crucial role in China's response to SARS.[8]

Mason argues that China's public health reform after SARS repositioned local public health in line with global health preparedness, resulting in a troubling "bifurcation of service and governance." "Local public health in China," she claims, "became geared toward the protection of global, rather than local, interests and toward the protection of a cosmopolitan middle-class dream rather than toward the betterment of the poor."[9] Mason shows that public health professionals increasingly justified their work in terms of what she calls the "common," "an idealized world of modernity, science, and trust" that linked workplace objectives and private aspirations in "an imaginary of the professional, the scientist, the cosmopolitan Chinese." In doing so, she argues, they also tended to neglect service and care for the populations that they governed, particularly certain marginalized sectors such as migrant laborers.[10]

In China's veterinary reform, much like Mason's description of public health reform, global preparedness programs are closely intertwined with trajectories of professional aspiration, as the response to the avian influenza crisis encourages the training and advancement of official veterinarians. However, Mason's critical argument must be situated in the somewhat different conceptual terms of the pandemic epicenter that I have developed in this book. Mason argues that biosecurity interventions supplant local needs (such as the "betterment of the poor") with global demands for pandemic preparedness. This opposition of the global to the local is a critical form widely used in anthropological studies of global health. In these studies the global is figured as a universalizing standard that disregards the particularities of a local site, whether that standard is a lifesaving humanitarian kit or a pandemic-preparedness plan.[11] Ethnographic accounts of these encounters tend to juxtapose the rich specificities of a local place to the abstractions of global forms, exposing what is missed in the gap between them.

But as I have argued throughout this book, the pandemic epicenter is not best understood as a local place. By following movements into the epicenter, this book has examined the intersections of global forms with several different strata across various spatial domains and scales, including ecological zones, working landscapes, and geopolitical territories. Rather than a bifurcation of global and local interests, therefore, I observed opposing trajectories of movement and the emergence of multiple new assemblages. On the one hand, training programs such as the FETPV heralded the movement of global health infrastructure and projects *into the epicenter* by transferring the techniques of epidemic field investigation to China's veterinarians. On the other hand, they also initiated a trajectory of professional *detachment* that separated state-employed vets—now remade as official veterinarians—*out from* the working landscapes of rural China, separating them from the duck farmers, egg traders, truck drivers, and all kinds of animals that inhabit these landscapes. Much like the laboratory model of expertise, adopting the new technical tools for investigating epidemiological objects demanded the ethical separation of veterinarians from broader society, a process I call detachment by professionalization.[12]

At the same time, the working landscapes that the official veterinarians left behind did not remain vacant. As the veterinary reform cultivated the quality of elite state-employed vets, the structural administrative reforms also reduced the quantity of base-level veterinary workers employed within the AHVS system. But when the curtailment of base-level AHVS veterinary services left a vacuum in veterinary expertise, a new kind of informal, unlicensed vet emerged at places like the Xiaolan poultry egg market near Poyang Lake, figures that I came to call "duck doctors." These practitioners are not the return of "traditional" knowledge specialists, suppressed by the state-led veterinary modernization of the 1950s. Indeed, duck doctors are innovative for their specialized focus on bird diseases and predominant use of industrial pharmaceuticals, such as antibiotics and vitamins. At the same time, their approach to veterinary work is very different from today's official veterinarian. The duck doctors who cluster around poultry egg markets and rural county roads, many of them only minimally trained, highlight the importance of an ethics of personal familiarity in their treatment practice, often prizing friendship with farmers and trust over specialized knowledge and technique.[13]

By looking back at the veterinary reform from Poyang Lake, one therefore observes something that looks more like stratification than bifurcation.[14] As official veterinarians are increasingly qualified in specialist knowledge, they are at the same time detached from the working landscapes of rural China. I observe this detachment in three domains: sociologically (as they become middle-class professionals), ethically (as they are authorized under new licensing regimes), and epistemologically (as they are trained in epidemiological thinking). But this chapter shows that veterinary reform is a process with two sides, driving *both* the detachment of the official veterinarian and the newfound attachments of the duck doctor. This double movement, observed synoptically, describes a process of stratification: a separation of veterinary practice into an upper layer—the state-authorized veterinary knowledge of the office veterinarian—and a lower layer—the personal familiarity with rural working landscapes of the duck doctor—a division into two strata of veterinary practice, increasingly impermeable and worlds apart.

COMMUNE AND CRISIS

The control of livestock epizootics was among the guidelines set out in the foundational common program (*gongtong gangling*) issued by the Communist government in 1949: "Protect and develop livestock industry, prevent and control animal epizootics." However, the government could rely on only about 1,800 educated and technically trained veterinary personnel in the entire country at the time, mostly holdovers from the upper administration or research institutes of the Republican government. At the rural village level, farmers relied on "folk" vets with no regulation by the state, including some farmers who conducted veterinary work on the side.[15]

During the 1950s the government began to consolidate, centralize, and extend veterinary administration, notably through the establishment of a livestock division (*xumuke*) in many county-level governments. By 1952, there were 967 county-level livestock divisions and around 10,000 veterinary personnel. A core Communist strategy during the 1950s was the unification and organization of folk vets through basic training programs, a strategy also used in human medical and public health sectors.[16] Several large-scale mass immunization campaigns took place during the early 1950s, initially focused on cattle diseases. By 1955, these campaigns successfully eradicated rinderpest and controlled foot-and-mouth disease, so the focus of campaigns shifted to the control of pig and poultry diseases, including a 1958 unified national campaign against Newcastle disease.[17]

The formation of people's communes in 1958, which grouped farmers into several production teams within a township-scale commune, greatly aided the extension of veterinary administration across the country. In 1959 the government established Animal Husbandry and Veterinary Stations in every commune, paralleling an earlier extension of anti-epidemic stations for human public health.[18] Each AHVS typically had three to five veterinary workers, with responsibilities ranging from vaccination to quarantine of infected farms and livestock breed improvement, and with a primary mandate for disease eradication.[19] The AHVS was administered according to what is called a *tiao tiao*, or vertical management structure, meaning that county-level plans allocated responsibilities and tasks to the AHVS.[20] The commune paid AHVS veterinary staff with work points, identical to the work points given for farmwork, and not from fees charged to farmers for services provided.[21] As a result, as Zheng and colleagues argue, the AHVS built strong linkages between the emerging nation-state and the local society, allowing the central state to directly mobilize and guide mass movements.[22] By 1967, nearly 40,000 commune-level AHVSs employed more than 200,000 professional veterinarians, and more than 500,000 additional farmers had been recruited as brigade-level antiepidemic personnel.[23]

By unifying folk veterinarians and farmers in state-managed stations, and providing brief training in veterinary techniques such as immunization, the AHVS invoked a characteristic Maoist configuration of anti-expert and antitraditional values. Like their more famous counterparts the barefoot doctors (*chijiao yisheng*), contemporary reports and other documents often refer to these rural veterinary workers as "barefoot veterinarians" (*chijiao shouyi*).[24] As Fang Xiaoping has argued, although barefoot doctors followed in the wake of Mao Zedong's critique of urban-focused professional health expertise, the barefoot doctors were also antitraditional because they introduced new biomedical treatment schemes and pharmaceuticals to rural areas. In the veterinary sector this extension of modern state veterinary services caused conflicts with existing animal healers and providers of livestock services. Sigrid Schmalzer, for instance, documents conflicts between a group of young women, trained in basic veterinary care at the county veterinary station in rural Jiangxi Province, and "folk vets" and "local boar keepers" in their home village. As Schmalzer summarizes, "Even as some people, including women, found opportunities to learn new skills and acquire scientific knowledge [by extension of state veterinary services], others found their livelihoods threatened."[25]

Beginning in the late 1970s, the initiation of administrative and market reforms to the collective economy transformed this veterinary system in two ways, a reconfiguration that Zheng and colleagues gloss with the phrase "heavy on husbandry, light on epidemic prevention."[26] On the one hand, economic reforms stimulated massive growth in livestock production, and on the other, administrative reforms led to the withdrawal of state funding for epizootic disease control.

I have previously discussed the market reforms of the livestock sector, but I will briefly revisit them here. First, collectively farmed land and livestock were redistributed or contracted to households under the "Household Responsibility System." By 1983, people's communes were disbanded. Rather than working collective fields in exchange for work points or ration slips, farmers now farmed their own plots, meeting production goals in the form of grain tax responsibilities. In their spare time, farmers could focus on sideline enterprises such as livestock husbandry, and sidelines grew to become the primary economic activity for some "specialized households." In addition, market mechanisms (including rural markets, legalization of long-distance trade, and the elimination of purchase coupons) gradually replaced the unified procurement and supply system of the planned economy. Government policies permitted market trading of livestock and poultry long before other agricultural products such as basic grains. Soon the growth of a manufactured feed industry, the introduction of new imported breeds, and the expansion of transnational agricultural capital intensified the market-driven growth in animal production, setting off China's livestock revolution. As the veterinary trainees explained to me, the livestock revolution meant not only increased numbers of animals, but also expanded transport of animals across the country and even internationally. Until the late 1970s, the husbandry system was "sealed up," "Teacher" Huang, a mentor and instructor from the CAHEC supporting the FETPV training, told me. "An animal born in one county would be raised, slaughtered, and eaten in that county. But today, a pig raised in Heilongjiang [the northernmost province] may be sold and eaten by someone in Guangzhou [a southern province]."

During the same period, China also undertook a related and equally dramatic reform of administrative governance. The government placed technical research and professional expertise at the basis of policy, a turn often described as a rebirth of scientific knowledge after the political excesses of the Cultural Revolution.[27] However, I learned from veterinarians that the impact of these reforms was much more ambiguous in the animal health sector. Teacher Huang told me that after the reforms the government did not properly care for veterinarians, particularly those trained in epidemic control. The number of university graduates in veterinary medicine did not increase, despite the massive growth in animal production and circulation. More troubling, perhaps, some veterinarians began to leave government service to start their own livestock enterprises or to work as animal health experts within private livestock companies, where pay was better. "In this way they could sell their knowledge for money," Zheng complained, with a hint of moral reprobation. Certainly, wages and status stagnated for many state-employed professionals during the reform era, including academics, medical doctors, and vets, while market opportunities grew.[28]

In rural areas the elimination of the people's commune structure effectively withdrew the primary source of funding for institutions such as Animal Husbandry and Veterinary Stations, just as it did for the Rural Cooperative Medical System and Anti-Epidemic Stations in the human health sector.[29] Although the state continued to pay part of the salary of vets, the AHVS increasingly financed itself by charging fees for vaccinations and other services.[30] For example, in one Jiangsu county, the veterinary stations shifted their primary work from epidemic prevention toward provision of technical support for livestock production, including the establishment of breeding farms and feed mills. This paralleled a shift from cooperative financing toward a fee-for-service, contract payment model.[31] Following these reforms, "Although village and township Animal Husbandry and Veterinary Stations were technically base-level stations of nation-state governance, the vast majority had become completely self-supporting."[32] During the 1990s, fiscal policy required the devolution of AHVS from county to township-level government, further exacerbating the funding crisis.[33]

Liang Ruihua, in a dissertation on avian flu control policies, offers a damning assessment of the consequences of 1980s administrative reforms:

> After the village administrative reforms, many town and village governments did not attach importance to veterinary epidemic prevention work, salaries for veterinary personnel were not included in the budget, the veterinary station had to support itself, veterinary personnel had relatively low salary benefits, and in addition it was not always released in a timely fashion, forcing many veterinary personnel to change professions. In some places, the veterinary station was contracted out to individuals; in other places, it was closed down completely. Some veterinary stations didn't even have an office building, its staff every year had to find "traveling (borrowed) rooms" in order to do their work; some animal husbandry and veterinary stations had leaks, walls falling down, broken windows. Until recently, across the country most veterinary stations lacked examination equipment, and had no laboratory machines, so in order to diagnose animal diseases the staff depended on their own eyes and experience, truly descending to a "one pair of scissors, one pair of eyes" condition.[34]

CULTIVATING QUALITY

Beginning in the early 2000s, leading veterinarians began to call for a new era of veterinary reform that would realign the structure of veterinary administration to reflect the transition from a planned to a market economy, including the requirements of free global trade that followed from membership in the WTO. In an article in the journal *Meat Hygiene*, Cui Yuying articulated the reform as a shift from a *guanjia* (literally, "imperial bureaucracy") to a *guanfang* ("official") veterinary administrative system. For Cui the guanjia system is fatally linked with the excessively centralized management of the planned economy. In the guanjia system, Cui argues, governance and business enterprise are united in the same institutions, and regulatory functions are not separated from technical support and extension services. Furthermore, administrative governance is divided across a staff of more than 500,000 people, primarily rural vets associated with the AHVS system.[35] As another commentator complains, most of these staff are "non-experts" and poor quality (*suzhi cha*), making "local protectionism" a serious problem.[36] In order to create a guanfang system, Cui calls for China's veterinary staff to be divided into two groups: one, smaller part, will undergo training and examination to competitively become "official veterinarians" (*guanfang shouyi*), upon which they will be paid by the state and will represent state authority in veterinary investigations and inspections, and the rest will become "professional vets" (*zhiye shouyi*), who will earn money from market-oriented services and will not represent state authority.[37]

The veterinary reform closely linked changes in administrative form with changes in professional ethos, reorganizing the structure of bureaucratic office along with official vocation. In 2004 MOA created a veterinary bureau and appointed a chief veterinary officer to represent China's veterinary sector in international diplomacy and trade negotiations. County-level governments also began to reform the Township Animal Health and Veterinary Stations, often by linking them with the County Veterinary Bureau and promoting the township veterinary workers to "official" status. In some areas, AHVSs were divided into three distinct agencies, with autonomous control over personnel, funds, and equipment: first, a smaller AHV station devoted primarily to livestock improvement and production growth; second, an animal epidemic disease control center devoted to disease surveillance and epidemic response; and third, an animal medicine and feed inspection office.[38]

Then, in 2005 and amid the height of the avian flu crisis, China's State Council, the country's highest governing body, issued a directive acknowledging and calling for extension of the national veterinary reform. The directive, titled "The State Council's Proposals on the Promotion of Veterinary Administration Reform," remarks on the ongoing establishment of an "official veterinary system," including the creation of the Veterinary Bureau. The State Council calls for the improvement, through training and licensing measures, of the new "official veterinarian" (*guanfang shouyi*): "An official veterinarian is a national state veterinary worker who holds recognized qualifications or license, legal authority or government appointment, and the power to issue health certificates. With reference toward the international common practice, the national state has begun to slowly institute an official veterinary system. Currently existing veterinary workers are required to improve all-around quality [*zonghe suzhi*] and professional level through professional training, and following recognized licensing or government appointment, to slowly enter the official veterinary staff."

The veterinary reform seeks to reduce the large quantity of veterinary workers in the AHVS system—many of whom are considered to be "nonexpert" and "low quality" (*suzhi cha*)—by using training, examination, and licensing procedures to select and improve the quality of "official veterinarians." Discourses of *suzhi* or "quality" are widespread in contemporary China, and as anthropologists have noted, the suzhi concept is powerful because it is relational. According to anthropologist Andrew Kipnis, suzhi "marks the hierarchical and moral distinction between the high and the low and its improvement is a mission of national importance."[39] In many cases, suzhi discourses are at stake in differentiating urban, middle-class elites from rural farmers and migrant workers. With roots in population policy and education reform, suzhi is contrasted with an image of vast numbers and rote education associated with the collectivist past: reduction in quantity is linked with improvement in quality. Therefore, the inclusion of veterinary reform and "official veterinarians" within suzhi discourse sets in motion several metrics of differentiation. First, the veterinary reform opposes the suzhi of the official veterinarian to the large cohort of minimally trained village and country vets, who are sometimes even directly referred to as "low quality." At the same time, the discourse intersects with veterinarians' own sense of being backward and behind in relation to the more cosmopolitan sectors of China's bureaucracy, such as human public health.

Across many domains, suzhi discourses contrast middle-class children, urban residents, and professionals (as sites or objects of suzhi training and accumulation) with rural bodies and forms of life (lacking suzhi, of "poor" quality).[40] Such a separation is easily configured, perhaps, in situations such as the building of gated communities, which are clearly and explicitly separated from rural environments and poor neighborhoods of the city. But the articulation of the veterinary reform as a process of suzhi improvement raised a specific tension. On the one hand, like their counterparts in human public health, the veterinary reform articulated the new "official veterinarian" as a renewed middle-class, cosmopolitan professional—one with technical expertise, state authority, international connections, and white-collar modes of work, what Katherine Mason calls the "common."[41] Yet at the same time the very object of veterinary work remained, unavoidably, rural working landscapes. How then would improvements of quality affect the relationship of veterinarians to farmers, farms, and farm animals? How would this changing social relationship affect the production of veterinary knowledge and the provision of veterinary governance?

In the next sections I focus on how this tension took shape during the field epidemiology training course. On the one hand, the content of the course advocated an epistemology of the field and a pedagogy of the particular. On the other hand, the form of the course followed a typical format of "trainings" (*peixun*) oriented toward the cultivation of suzhi: the recruitment of trainees from the provinces to centers such as Beijing or Qingdao, perks such as free computers, periodic examinations, contact with international trainers, and certificates of completion. Although the training course introduced the concepts of *field* epidemiology, the course also exemplified a sociological process of professionalization that took the trainees farther and farther from the forms of life and landscapes of rural China. Just as the field was coming into view as an epistemological object, the social ties to the working landscapes where these field sites are located were broken.

FIELD TRAININGS

The FAO Emergency Center designed the FETPV training course to bring China's official veterinarians into the field. As Vincent Martin and David Castellan had stated in the proposal to the MOA, the first module would focus on the application of epidemiological concepts, disease surveillance, and outbreak investigation "under a wide variety of field conditions."[42] During the course I observed that the lessons used techniques of simulation to progressively move concepts from a classroom setting into field applications. First, a concept was introduced through a short lecture (such as the statistical concepts of rate and ratio). Then, beginning with the virtual space of a computer game, these concepts were brought to imaginary field settings through problem-based exercise on fictional outbreaks. Next, we moved out of the classroom to mock outbreak investigations on Beijing farms. In the final stage of simulation, drawing from the model of "training through service" that is part of the CDC's standard international FETP course, the trainees conducted actual epidemiological investigations under the auspices of their home offices and submitted the findings to the FETPV instructors for correction, improvement, and assessment.

In the process the field was reconstructed as an epidemiological object—that is, using techniques of statistics, the trainees turned working landscapes into numerical populations. "Field" epidemiology did not simply mean on-site visits to farms, but rather a new manner of assessing risk and determining appropriate interventions on farms. As the instructor Dirk Pfieffer told the trainees, describing his own work on bovine tuberculosis in badgers, "Every time I go to a farm that has TB, badgers also have TB. But this is only an anecdote. It is meaningless." Only by collecting observations into numerical populations and assessing them statistically can their true meaning be understood. This statistical accountability of veterinary work was very different from the substantive rationality of prevention that typified veterinary governance in China, exemplified in programs such as the universal vaccination of poultry, in which goals were set without numerical calculation of risks or costs.[43] Throughout the course I observed the trainees negotiating the remediation of their relationship to rural farms, animals, and farmers demanded by statistical styles of thinking.[44]

During the second week of the course we woke early to board a bus out to the outskirts of Beijing, where we visited a number of large-scale farms in order to conduct mock outbreak investigations. Suiting up in white lab coats and covering our shoes with blue plastic booties, we interviewed the managers of one poultry and two pig farms, following questionnaires developed in the classroom the previous day (see figure 6.1). Although we visited actual farms, in the process the farm was reconstituted as an epidemiological field site through the use of a formal survey instrument. The veterinarians were very familiar with farms and often joked with the farm managers, particularly about the unusually high quality of these farms, which it turned out were actually Beijing government "model" farms. "These are probably the two best farms in Beijing," one trainee commented with a smile. Certainly, the trainees noted, these farms did not represent the rural countryside as they understood it.

All of this easygoing conversation posed some problems for the trainees. They struggled to fit the farm managers' wide-ranging commentary into the confines of the questionnaire, often writing in the margins or even flipping the paper over to add notes on the back (figure 6.2). At this point the FETPV instructors reminded them that most of this familiar conversation was extraneous to the outbreak investigation; they should be consistently administering the survey questionnaire in order to produce the kind of data that could be statistically assessed. At most, one could include a blank space on the survey instrument in the future for free-text commentary. Field epidemiology and statistical surveys were much more than a new set of techniques that the veterinarian could adopt, I began to realize: as a distinct style of reasoning, it required adopting new ways of assessing the true and the false. I recalled that as we were designing the survey questionnaires the day before, one trainee had

FIGURE 6.1. Veterinary trainees visiting a chicken farm to conduct a mock survey.

asked the instructors how questioners could ensure that farm managers were telling the truth. The instructors rephrased this concern as "the problem of data gathering," "how does one word a question in order to get reliable results," and "quality control." Moreover, they highlighted how statistical tools, such as confidence intervals, could be used to assess the reliability of results. But the trainee, I knew, was asking a different question: what happens if people answering the survey questionnaire are simply lying? At the farms the trainee veterinarians had tried to develop rapport with farm managers, hoping that the cultivation of personal familiarity could provide a foundation for honest speech. The statistical tools provided a completely different approach to validation and, along with it, required the veterinary trainees to cultivate a different ethos of engagement with farmers and farm managers.

The effects of the FETPV course were not limited to the transmission and internalization of epidemiological techniques. Unavoidably, the FETPV course also exemplified the form of a *peixun*, or training course, and therefore sociologically fit within what Jie Yang calls the "mushrooming *peixun* culture" in contemporary China.[45] Training courses are booming in almost every sector of Chinese society, as documented by anthropologists in domains as diverse as psychotherapy and reemployment services.[46] Yang points out that peixun

FIGURE 6.2. Conducting a survey. Farmer at center filling in form, with attentive veterinary trainees assisting.

is more than simply an educational course, for the term includes a cluster of meanings that can be "roughly translated as training, teaching, or cultivating." By linking education with ethical cultivation, peixun is one of the key mechanisms of individual self-improvement in the pursuit of suzhi quality. As Yang puts it, the content of peixun courses tends to offer "practical knowledge or skills that give people an advantage in the job market."[47] However, in many professional or state employment settings, the peixun form itself, rather than the specific content of the course, is also an ethical mechanism of self-improvement. This is because participation in peixun courses involves taking a holiday from work, or a retreat; traveling away from the office to other parts of the city, region, or country; receiving gifts or perks; and, of course, banquet dinners and karaoke sessions with fellow trainees. Moreover, the very ubiquity or overabundance of peixun trainings means that for many participants the organizational means may be more valued than the presumed ends of the training course.

The trainee veterinarians stayed in a standard but modest hotel across a busy intersection from the Hygiene Inspection Service offices where they attended the course lessons. A scrolling red ticker-tape screen at the entrance to the hotel welcomed "participants in the first China-FETPV" and also welcomed an accountants group for their own training course. In coffee breaks and other time we spent together outside of the classroom, trainees constantly discussed what aspects of their participation in the peixun would be reimbursed by their office, such as travel to Beijing, daily transport, meals, or other expenses. These conversations were always comparative, detailing the differences between *danwei* offices: more or less claimable expenses, higher or lower official status within the state bureaucracy, where office funds could be turned to private use, and where the new anticorruption surveillance measures made careful accounting necessary.

One day when we left the classroom for lunch, Zheng, a trainee who also worked at the Hygiene Inspection Office, pointed out the three vehicles—station wagons and SUVs—parked in the office lot. The trainees began comparing whose danwei office had more cars and whether leaders could drive office-owned cars for their personal use or strictly when conducting office business. Perhaps unsurprisingly, a trainee from a national-level office located in Beijing noted that the leaders in her office were not allowed to have cars, whereas a trainee from a Beijing city-level office laughed and said that the leaders in his office had seven cars at their disposal, several with private drivers. Hua Weinong, a trainee from the Guangxi Zhuang Autonomous Region CADC, then boasted that his danwei had paid for him to take English lessons with an American teacher and, not only that, also for a car to drive him there and back. The other trainees laughed: "Your danwei must be so rich!" Hua, a bit abashed, dissented. "No, no, no," he said, waving his hands, "my danwei is not rich." But because Guangxi is considered a likely epicenter of zoonotic pandemics such as influenza, he explained, a lot of international organizations conduct investigations there. As part of these projects, they provide funds for training and equipment to his office—and even English lessons. When I visited Hua at his office in Nanning several months later, he shared a small, cluttered table with another colleague, each working on an old PC. Across the danwei courtyard, however, Hua showed me a glistening new six-story office building. Although his own desk in the new building was not quite ready, someone was already working at the bench of the laboratory facilities on the first floor when we walked in to take a look. The smell of fresh white paint was almost overpowering.

THE OFFICE AND THE FIELD

During the third week of the training course, Xu, a trainee from the Beijing Animal Center for Disease Control, stood at the chalkboard to present his "study design" to investigate Newcastle disease. Xu explained that it was a study of prevalence (number of cases of disease at a given time divided by number of people at risk) and outlined the steps by which he would collect the appropriate data. At this point, a trainee from Chongqing asked him what the hypothesis of the study was. The instructor agreed: the study is perfect, but there is no hypothesis.

"There is no need for a hypothesis," Xu replied calmly.

"Of course you need a hypothesis," the instructor responded. "Because depending on what hypothesis you are trying to prove correct, you will need a different sample size. If you really have no idea what the prevalence is, you still have to have a hypothesis, so you will hypothesize 50 percent. Moreover, depending on the hypothesized prevalence, the sample size will vary, and therefore the cost of the study will vary. So when you ask the government official to fund your study, you can tell them exactly how large a sample size you need to verify your hypothesis. Otherwise you may be just wasting money."

Xu smiled. "But things don't work that way in China! You don't ask for money; you are simply given a certain amount of money to conduct a study."

Someone else agreed. "The leaders (*lingdao*) don't understand these scientific concepts. Telling them the 'sample size' you need is just a waste of words and paper."

Whether or not the trainees would be able to actually use the field epidemiology they learned once they returned to their offices emerged as a core concern among organizers of the training course. In referring to "organizers" I primarily mean the national staff of the Emergency Center (including Cai Haifeng), and the "mentors" from China's CAHEC (including Teacher Zheng), who were responsible for the day-to-day management of the course, rather than the FAO officers such as Vincent Martin who planned and initiated it. As we will see, organizers like Cai and Zheng worried that the technical expertise and field-based thinking that formed the content of the course could easily be lost in the bureaucratic formalism and hierarchical management of China's state veterinary offices.

Their concerns about the relation between science and the state in modern China resonate with a long-standing academic literature. The historiography of science in modern China, notes historian Zuoyue Wang, has been dominated by concerns about the role of the state. Until very recently, most English-language historical narratives were rooted in a diagnostic pathology of social structure, examining the forms of politics, social order, and the state for evidence of distortions or obstructions to valid scientific inquiry.[48] As Wang notes, this historiography is grounded in Robert Merton's sociology of science, and in particular its argument that the ethos of modern science is promoted by democratic societies and constricted by authoritarian ones. As a result, these historical accounts are constantly searching for the "self-contained authority of a cosmopolitan science community" to emerge from beneath the oppression or influence of state power, as Laurence Schneider once put it.[49]

For the organizers of the field epidemiology course, the cultivation of "official veterinarians" (*guanfang shouyi*) was also fraught with the relationship between science and state. However, their concerns were not about state control or censorship of free inquiry, nor did they suggest that this had anything to do with the presence or absence of political democracy. Rather, their more precise and pragmatic warnings articulated a possible tension between the on-the-ground methods of field epidemiology and the formal systems and hierarchical structures of the bureaucratic office. In wider debates, complaints have emerged that many official veterinarians are more officials than veterinarians; for example, being designated an official veterinarian depends on successfully passing civil service examinations, which test for political knowledge and loyalty rather than technical expertise. In the case of the FETPV course the organizers worried that the social form of the office—and in particular the hierarchical authority of the leader—would overwhelm the epistemological content of the training course. Would the trainees really become a new kind of field epidemiologist, able to remediate field situations through statistical concepts such as population? Or would the course be just another licensing mechanism, another rubber stamp needed for advance on the ladder of the bureaucratic career, a new ready-made entry on the curriculum vitae, with the important technical content soon forgotten once certification was in hand?[50]

Teacher Zheng raised these concerns with the trainees one evening sometime after Xu's presentation. Zheng was one of the more influential mentors. Short but stocky, clean-cut, he was always energetic. He often sat at the back of the room, translating lectures by the international trainers into Chinese for anyone sitting nearby him. Sometimes, frustrated by the apparent lack of understanding among the trainees, he strode to the front of the classroom, taking the podium from the international trainer and going back over the lecture in Chinese, often adding his own insights. On this evening, we had returned to the conference room after a meal in the dining hall of the nearby hotel. The international trainers were either out to dinner or back in their rooms resting. A few trainees had cups of coffee, and there was a warm feeling in the room as they worked (or instant-messaged friends and family) on their laptops. Zheng began to tell a story, and everyone looked up, listening intently. Back in 2008, he said,

China was trying to prove that there was no avian influenza in the country because they wanted to force the United States to lift a ban on chicken imports. Chinese officials presented a study, a *beautiful* study, describing laboratory results that showed no positive results. An American official asked them: what was the sensitivity of the test you used? The Chinese leader presenting the study simply stared straight ahead for a moment, silent, then bent over to his colleagues and muttered: *what the hell is sensitivity?* They explained the concept to him, this way and that way, until finally he straightened up, adjusted his tie, and said to the audience: I guess it's about 95 percent.

Using this story as background, Zheng redirected the focus of ethical concern from the leader to the veterinary trainee. Describing the techniques for conducting outbreak investigations, he asked the trainees: "When does one conduct an outbreak investigation?" He then answered his own question: "When the leader says go, you just go; this is also incorrect. In order to conduct a systematic investigation, a lot of preparation work must be done before heading to the field. This doesn't mean to *oppose* the leader but to take the leader's order and make it scientific research."

At the very least, the trainees struggled with balancing demands made by their office leader with their participation in the course. Often it meant working overtime. One trainee from the Beijing Animal Hygiene Inspection Service was particularly harried. Because the first module was held in the office building of his danwei, he usually attended the training program during the day and worked overtime at night. Sometimes he hardly slept. Precisely because of these challenges, the second training module was located in Qingdao, giving the Beijing-based trainees a modicum of independence from the demands of their leaders.

Another incident reveals the challenges the trainees experienced in negotiating obligations to the training and to the lingdao. Deng is a pathogens expert at the Key Laboratory for Avian Influenza in Harbin, a city about twenty hours by fast train and several hours by plane north of Beijing. During one part of the training, he struggled to both accomplish the requests of his lingdao and fulfill his work in the training program. One week he left early on a Friday morning, returning to Harbin for two days, then returned to Beijing for class on Monday. The following weekend he again flew back to Harbin before flying to the southern province of Jiangsu to collect samples from poultry, finally returning to Beijing on Tuesday. Anticipating the travel, he complained that he had submitted the proposal for the study months ago, but his lingdao only that day replied and demanded that the study go forward. "There's nothing I can do about it; lingdao are like this," Deng told an FAO staff member and myself in a low whisper, while a lecture was in progress. "Next year there are next year's things."

As it turned out, the first module of the FETPV took place at a particularly inconvenient moment in the cycle of Chinese bureaucratic time. Held in December 2010, it happened to be just more than a month before the end of a five-year plan. All offices throughout the Chinese bureaucracy were busy compiling the paperwork necessary for final reports to be submitted to the higher authorities, accounts of the past five years and proposals for the next. All of these reports were to be finalized and submitted by the onset of the Chinese New Year holiday in February. Research needed to be completed: the more results the better. The veterinary trainees were caught up in this flurry of paper. If they weren't working double time during the course itself, a growing pile of papers was waiting for them on their desk when they got back to their danwei in early January.

Although for the trainees this was experienced as a very immediate challenge of time management, the organizers felt that deeper concerns about the future of China's official veterinarians were at stake. One day, as we sat at a common lunch between classes, Cai Haifeng attempted to personally persuade the veterinarians to change their engagement and attitude toward the training. Cai spoke quickly and curtly, with disgruntled animation. "Is everyone understanding the lectures?" he asked. "Are there problems with language or accent?" All of the trainees answered that it was better, it was better now, things were getting better. A few halfhearted jokes were made about the difficult accents of the trainers, two of whom spoke English with German or French inflection. Cai looked up from his noodles to meet the eyes of his audience. "This is not like other trainings. It must be taken very seriously," he said firmly. He noted that the FETPV is based on a "training-through-service" methodology, which means that the concepts and techniques they learn should be put to use in their everyday work. In fact, he reminded them, one of the requirements of the training would be to conduct a long-term epidemiological research investigation as part of the work at their home danwei.

A woman from the China ACDC, based in Beijing, spoke up. "How will we be able to do these research projects when we go back to our danwei? We are already taking a month off, and won't we have to return to regular work when we get back?"

Cai's face tightened. The FAO already made agreements with each of the danwei, he told the trainees. It had all been made very clear: the danwei must allow them to conduct these research projects. It was very important that the training not go to waste.

Leaning forward in his chair, Cai began to relate a story about his own experience in a training program during the 1990s. At the time he was a young veterinarian working at the Qingdao Animal Quarantine Service, now renamed as the China Animal Health and Epidemiology Center. The Ministry of Health invited veterinarians to a training program to learn techniques of epidemiological research. The MOH spent a lot of money on the training. But when Cai returned to work, he was unable to put any of the things he had learned to use. Soon he forgot his new skills. "Now, just this month-long module of the two year FETPV program alone is costing over one million RMB [around $150,000 at the time]. Let's not let it go to waste," Cai concluded.

The problematic ethos of the trainees, their lack of attention or commitment, was also raised in official plans for the program. The meeting minutes from the 5th Steering Committee Meeting for the program, held in January 2012 as the first cohort of trainees was nearing completion of the course, report that Dr. Sun Yan, head of the Division of Science and Technology and International Cooperation, Veterinary Bureau, commented that the FETPV training needed to "better motivate/push the trainees to be more committed to the training."[51]

VETERINARY CAREERS

In private, Cai told me that he worried that the "management system" of the veterinary administration would waste the technical knowledge in field epidemiology provided by the training course. He complained that Chinese official veterinarians prioritize pleasing their superiors over the conduct of good field investigations. "They don't care about society; they care about the higher-level leaders," Cai said. "They don't care about the common people; they only care about the higher-level." However, as I got to know Cai better and he told me more about his own career, I began to sense a second, somewhat different, diagnosis of China's veterinary crisis: rather than an opposition between official management demands and the pursuit of scientific knowledge, Cai described a process of scientific detachment from rural landscapes, a process in which certain forms of professional expertise produced their own myopias.

Cai came from a poor background in Shandong Province. His college degree was paid for by the government on the condition that he would work in government service after graduation. In the late 1980s he got his first job at the animal quarantine service in Qingdao, Shandong Province, the main national institution devoted to control of epidemic disease in animals. Within a few years, however, Cai found himself working amid the increasing pathologies of an unprecedented livestock revolution. According to Cai, it was not until the 1990s, "very late" (by international standards), that the large-scale broiler chicken enterprises took off. Many of these large farms were established in Shandong.

In the past, Cai explained, Chinese farmers had raised chickens in a "backyard" manner: small flocks raised in family courtyards and considered a part of the household unit. All of a sudden, there were farms raising one hundred thousand, five hundred thousand, even a million birds. The rapid increase in poultry numbers and density caused many disease outbreaks: Newcastle disease, parasites, and "diseases you don't have in Europe and America." Some of the farms experienced average annual death rates from disease as high as 10 percent.

In Cai's opinion the increase in livestock, in living beings and their ecology of pathogens, was not in itself the reason for this increase in diseases. Rather, diseases appeared because of a failure of management and biological security. The model of large-scale farms was imported from Europe and America, but Chinese farmers had no experience in husbandry at this scale. "They didn't know how to raise chickens on a large scale. Nobody did; I didn't know how."

Cai was often assigned to investigate and control outbreaks on large-scale poultry farms. The visceral experience of sick and dying fowl; the red, white, green, and yellow excretions; and the pathological appearance of diseased internal organs developed into a long-standing repulsion: to this day, almost twenty years later, Cai will not eat chicken. Therefore, I was surprised that his reflections on that era are fond. He prized that time in his career, he told me, because his knowledge of veterinary care blossomed in close adjacency to the practices of production.

Cai was able to conduct full investigations on his own, from the survey of farm conditions to the laboratory diagnosis of disease, sharing experience with farmers and even developing recommendations for control interventions and how to prevent future outbreaks. Each investigation was distinctive. Cai did not have a standard protocol but rather cultivated a nose for the unexpected, searching out the trail of contagion.

One day, he told me, he had struck upon a perplexing outbreak of a contagious disease. When he arrived to inspect the farm, everything looked clean, bright, and orderly. Appropriate segregation and isolation measures had been instituted throughout the farm. Farmers lived in quarters separated by a good distance from the birds, the chicken sheds were carefully walled off from one another, access was restricted, yet disease still kept arriving at the farm. As Cai was strolling around, looking for any sign of unreported mixture or contact, a truck arrived at the gate. A few workers jumped out, entered one of the chicken sheds, and as the truck backed close to the shed, they reemerged carrying shovels full of manure, tossing the contents onto the back of the truck. With a few questions, Cai discovered that the truck traveled from farm to farm throughout the region collecting manure to repurpose and sell as fertilizer. It could easily carry pathogens along with the rich cargo from one farm to the next.

During this period, Cai skirted the lines between roles as an employee of the quarantine bureau and as a private individual providing veterinary services to the farmers. He told me that the large broiler companies, in particular, were suspicious of the government, fearful of quarantines or other disruptions to their commerce, so they "never told the government anything." I expressed surprise that the companies spoke to Cai openly: he was a government worker. Cai explained to me that he had made clear to the industrial farms that he would not report what he saw to the quarantine authority. At first, the farmers did not believe him, but over time they came to trust his word. They paid him for his work, and he earned much more from these investigations than from his regular salary. When Cai spoke with me, he emphasized that this situation was far from ideal. Yet the benefits he saw in this work were not only personal monetary compensation. There were also benefits for epidemic control, for China's rural economy, and, perhaps, even for truth itself. By developing trust with the owners and employees of the company farms, by breaking his own duty as an officer of the quarantine bureau by refusing to report the disease outbreaks, he was able to bring the actual materiality of farm production under the observation of scientific inquiry.

When in later years, just after the turn of the millennium, Cai advanced to more formally prestigious positions within the Ministry of Agriculture, he was brought farther and farther from the farm, and he sorely felt the pathos of this advance. His life experience, narrated as an increasing separation of knowledge from its object, embodied a diagnosis of the problems afflicting veterinary practice in contemporary China. He noted skeptically the recent increase of "professors" and lamented that students today favored laboratory and specialized topics such as molecular biology. Only in such specialized sciences could you "write a good thesis," he told me, yet he felt that these specialized knowledges were of little use in the practice of livestock disease control.

The veterinary reform, as I discussed above, intended to distinguish and stratify China's veterinarians into two types: state-employed official veterinarians and licensed but private "professional" veterinarians (*zhiye shouyi*). Cai argued that, in fact, the reform was actually producing a different typology: those trained in specialized sciences that have limited practical applicability and those who lack knowledge altogether.

During the FETPV training program, talk among the organizers and trainees frequently returned to the topic of veterinary reform (*shouyi gaige*). Whether hopeful or pessimistic about the future of China's veterinarians, agreement is widespread about its direction: toward the international standard, highly specialized, expert scientific, normatively professional "official veterinarian." In this way the veterinarian would not only meet international standards (*yu guoji xiangtong*), as one text on the reform puts it, but would also join the more advanced sectors of China, fulfilling the idealized transition from planned to market economy and toward scientific modernity. For Cai, however, the expansion of veterinary knowledge through international-led trainings and the veterinary reform threatened to have an ambivalent effect: by cultivating official veterinarians as subjects of increasingly specialized and elite forms of scientific knowledge, the possibility of sharing a social world or way of life with poultry farmers would become more difficult. Precisely as communication with the international scientific community intensified through journal publications and outbreak reporting, the lines of communication drawn by Cai between government science and poultry production were rendered sociologically unlikely, epistemologically invalid, and morally problematic, even corrupt. Yet the possibility remained that despite all of the survey instruments, virus sampling schemes, and computerized epidemiological models, the newly formed official veterinarians no longer really knew what was happening on the farms. Certainly, as one trainee had recognized, there was a good possibility that farmers might not tell them the truth. Although the epistemology of field epidemiology detached scientific facts from rural China's working landscapes, turning farms and livestock into statistical populations, this was not what worried Cai. Rather, Cai wondered whether epidemiological training, as one moment in the cultivation of a high-quality (suzhi) official veterinary career, would detach the vets themselves from the social and ecological worlds of rural working landscapes. This was an ethical rather than epistemological, and subjective rather than objective, detachment.

THE DUCK DOCTOR'S DIAGNOSIS

Although the keystone of veterinary reform involves training courses such as the FETPV, the most significant impact of the reform has been an enormous reduction of state-employed veterinary workers at the village and township level.[52] These two processes of detachment are closely interlinked. As I noted above, the large number of base-level state-employed veterinarians was one of the crucial problems identified in reform policies. In language resembling China's famous "One Child" population policy, government discourses on veterinary reform argue that it is necessary to reduce the quantity of vets in order to improve their quality (suzhi).[53] However, as the state has withdrawn from the local veterinary sector, and poultry production has exploded, informal animal healers have appeared offering for-profit treatments for animal diseases, including pharmaceuticals. As Zheng Hong'e and colleagues point out, "As soon as poultry fall sick, farmers will go to private veterinarians to seek assistance, and will not seek assistance from village or township animal veterinary officials."[54]

Out in the Poyang Lake region, I found that the government-employed "official veterinarian" was a minority among the duck doctors. Around the Xiaolan wholesale egg market south of Nanchang, the outer walls were lined with ten or so small livestock-medicine shops. More than twenty clustered on nearby roadways. The shelves inside these shops were filled with brightly colored boxes and packets of medicine, produced by factories in Shanghai, Chengdu, and Guangzhou. When duck farmers hired trucks to deliver eggs to the market, many also stopped to purchase medicines and preventative supplements, such as vitamins. And when ducks fell sick, although it was possible to bring the sick birds for examination at the veterinary station, most farmers preferred to seek assistance from the duck doctors like those at the Xiaolan market.

These shopkeepers are much more than pillbox salesmen. They earn their living diagnosing diseases, prescribing medicines, and providing advice to breeders—by curing ducks. Just outside a small shop on the outer wall of the Xiaolan market stood a small concrete block, rising to just below the waist of an average person. On one visit to the market, I stopped to watch as a farmer walked up the two concrete steps to this duck doctor's shop, a ten-gallon black plastic bag slung over one shoulder. Greeting the doctor, she dumped her bag, five or six dead ducks, gray and damp, feathers matted, falling to the ground. The doctor sprang to work. Moving quickly and with certainty, displaying a focused confidence in the movement of his limbs, he flung a duck carcass on the concrete block. He reached for a sharp blade and made a long incision from mouth to anus, then pulled open the body cavity. He thrust in his bare hands and sought the entrails. A string of coiled intestines in his grasp, he again drew the knife, slowly this time, incising along the length of the digestive tract. He peered intently at the contents exposed, rubbed them in his hands, and looked again. After a minute of inspection, the doctor raised his head and tossed the

FIGURE 6.3. Hao Weidong's medicines.

duck off the pillar to the ground below. He grabbed a second duck and went through the same steps, but more quickly, verifying his initial perception. After a third duck was tossed to the growing pile on the far side of the block, the doctor straightened up, rinsed his hands with a pail drawn from a bucket of water, and gestured for the breeder to join him in his shop. Once inside, he decanted a dark brown-red fluid from a large glass container into a small plastic bottle. With a few words of explanation, he handed the bottle to the breeder, who gave him a few paper bills in exchange.

Hao Weidong, whom I came to know best, owned one of the shops inside the marketplace. His shop was little more than a hastily constructed wooden shed squeezed between two of the large cages where egg buyers received and stored their wares. Shelves packed full with brightly colored packets of medicine, gold and orange and blue, lined the three inner walls (see figure 6.3). Hao proudly pointed out to me the plentiful "Western" medicines (*xiyao*) filling the shop. In a back room, he also stored a wide collection of Chinese medicines (*zhongcaoyao*), kept out of sight because they lacked the labels required by government inspectors. A strong metal gate blocked off the front of the shop when Hao was not there. A simple sign above the gate read "Duck and quail medicines."

FIGURE 6.4. Hao Weidong's shop.

The duck doctors opened their shops each morning when farmers began to arrive in shared trucks loaded with eggs. In the morning, Hao opened the shop by pulling up the metal grate, then moving a wooden desk to just out front. The desk, decorated with a few boxes of medicine and lined on both sides with boxes of product, formed a simple counter (see figure 6.4). After farmers sold their eggs, they typically went to purchase feeds and other supplies they might need, including medicines. Hao often stood on one side of the desk and greeted farmers as they passed the shop, offering cigarettes, asking about their birds, and inviting them to sit on one of the long wooden benches to talk for a while about ducks and quail and medicines.

Hao aimed to cultivate relationships with farmers not only as individuals but also in order to gain access *through them* to relationships with others. Any individual farmer also always represents a position within a web of kinships or friendships with other farmers. Hao made the importance of these webs of relations clear to me one day when he talked about the trucks that farmers drive to market. He works particularly hard to cultivate friendships with the drivers of trucks, giving small gifts or favors, or inviting them to eat in a "beef specialties" restaurant that is attached to the Xiaolan marketplace. Hao told me that farmers will often follow the recommendations of their driver when purchasing medicines. If the driver is a friend of Hao's, they will visit his shop instead of another.

This interest in trucks and their drivers is built on an understanding that the truck is a metonym for a web of relationships. The farmers who band together to lease a truck do so based on ties of familiarity, from either common membership in a village or literal kinship. Hao's strategy reflects the fact that a friendship with one member of a group that leases a particular truck almost inevitably brings with it a relationship to the other members of the group. The cultivation of this friendship with one member quickly flows through and extends into these other relations. A friendship with the truck driver, in particular, can lead to the development of friendships with the other members of the truck group.

On any given day the two wooden benches in the shop hold a revolving set of bodies: Hao's assistant, a woman whose family has connections to the management of the marketplace; farmers passing by or old customers, as well as their companions; and representatives from the pharmaceutical factories. That morning, Hao gave me a bottle of water and told me to make myself comfortable while he stepped out to run a brief errand. I watched the trucks pull in and out of the market, and watched the duck breeders unloading boxes of eggs.

Only a few trucks remained in the market by the time Hao returned. He carried a large black plastic bag with him and put it in the back end of the shop. At the back of the shop was an open doorway leading to a small back storage area. Hao ducked into the darkness and returned with a kilogram scale. Untying the large plastic bag, he pulled out six small pink plastic bags that exhaled a bittersweet smell. Roots and grasses and herbs: Chinese medicines. Looking at a prescription scrawled on a small piece of paper, he asked for his calculator and made a few calculations. He began to untie one small bag after another, pouring a percentage of their contents—dark gnarled roots, dried and twisted—onto the kilogram scale. He was preparing medicine, he told me. Satisfied with the weights, he then formed two identical piles on simple cardboard sheets. He carefully poured these pharmaceutical mixtures in two new plastic bags. Finally, he pulled some white bottles of factory-made medicine off the shelves and, I noted with surprise, placed them in only one of the plastic bags but not the other.

"A farmer had eight quail die in only two days," he explained. Just then, one of his mobile phones rang. I heard a woman's voice on the other end. "Hello, little sister," Hao said familiarly. "Yes, I am preparing some *zhongcaoyao*, along with some antivirals and *wei* C [vitamin C]." I expected this woman was the farmer with the eight quail but soon found that the relationship of diagnosis and exchange was more complex.

Hao and I set out with the two plastic bags of prepared medicines, his prescription given substance and weight. At a busy throughway Hao hailed the public bus. We rode for some time, passing through the urban center of the county seat, then out into open country. The bus came to a stop, the terminal station, in an old town that straddled a river, the buildings rebuilt in past years with concrete and white tile. Hao waved to a young woman on a motorbike across the street.

Leaving one of the black plastic bags with me, he walked toward the woman, and they greeted each other with bright smiles. After a short conversation, Hao handed her the bag of medicine. When she tried to hand him some cash, however, he vigorously refused.

"Why didn't you accept her money?" I asked when Hao returned to where I was standing.

"She introduced me to her uncle!" Hao said with a laugh. "I earn his money this time."

We walked across a bridge over the river to a small restaurant where the uncle, Old Wu, was waiting. Wu wore a lightweight blue work apron common to many of the rural farmers in Jiangxi. Inside the restaurant, Hao unpacked the medicines in the bag, explaining how to feed them to the birds. There are three packs of wei C, used over two days, or you could stretch it and use one per day. Suddenly, Wu pointed to the Chinese medicinal herbs.

"These should all be very cheap," he stated bluntly, suggesting that Hao's prices were too high. He explained that until retiring in recent years he had worked at the local Chinese medicine clinic and knew all of the herbal medicines and their value.

Hao, surprised, excused his prices. "It's the Western medicine that's expensive," he said, pointing to the white bottles.

After some deliberations, the two men decided that Hao would go to Wu's home to look at the quail (*kan anchun*), despite the fact that Hao had already prepared a prescription based on the symptoms reported to him over the phone. We rode in a three-wheel motor-powered cart out into the open countryside, which was dotted here and there with duck sheds and fishponds, crossed by train tracks and by power lines overhead. In the river a number of barges were mining sand, large cranes carving out the riverbed and building tall white mountains on the shore. Arriving in a densely built village, where a second story had been added to nearly every home in recent years, Wu led us into his courtyard. We walked into the house, then went up a small flight of stairs to the second floor, where Wu kept his quail. Hao commented on how clean Wu kept his quail cages, which usually produce a very strong smell.

FIGURE 6.5. Hao Weidong looking at quail.

Inside the upstairs room the walls were lined with tall wooden cages, and each cage was filled with the small squawking birds (see figure 6.5). Hao looked around, scanning the cages, and declared that the birds seemed to lack spirit. Wondering aloud about the temperature, he checked a thermometer on the wall. It read 28 degrees Celsius. Too hot. Perhaps this was the cause of their illness. After a few more words of advice, Hao protested that we needed to return to the shop, and in a few moments we were back in the three-wheel and headed back to Xiaolan.

DETACHMENT AND ATTACHMENT

To understand this diagnostic and treatment practice, it is useful to compare the contemporary duck doctor to the Mao-era barefoot veterinarians working at the AHV stations. First, Hao Weidong, like barefoot veterinarians, had obtained minimal training in veterinary medicine. "When I started," he told me, "I didn't understand anything, even about the Western medicines. I took a one-month training [provided by a pharmaceutical company] and then entered the market. When I started, I only carried a book bag," he said, pointing to his shoulder. "Inside the bag, I didn't bring any medicine; I only brought some brochures [advertising the pharmaceutical company's products]. If the farmer wanted a kind of medicine, I would write an X-mark and order it from the factory." He gave most of these medicines to the farmers for free, hoping that if they worked, the farmers would buy more. And perhaps more importantly, they would introduce him to their friends.

Also like the barefoot veterinarian, the duck doctor is not a *traditional* figure; he is not a "folk vet," using age-old methods, but uses cutting-edge pharmaceutical treatments for duck and poultry diseases, including antibiotics and antivirals. Many, like Hao Weidong, entered into doctoring later in life as a second or third career. Several of the diseases they treat, such as avian influenza, are widely understood by farmers to be novel and even unprecedented challenges. As a result, the duck doctors often develop innovative treatments: Hao often boasted of a cure for avian influenza that he had invented by mixing herbs and Western pills.[55]

Finally, the duck doctor is also like the barefoot veterinarian because his practice is rooted in the close social relationships he has with farmers. The duck doctor shares the same social world with them. For example, Hao Weidong will often combine diagnostic or treatment visits to farms with meals at the farmhouse, and sometimes invites his clients to meals at a special beef restaurant on the outside of the Xiaolan market. In at least one instance a client invited him to his daughter's wedding. Sometimes Hao even played the role of a therapist, listening to long confessions of a farmer's troubles that had nothing to do with duck farming at all, and offering suggestions and warm advice.

Yet, despite all of these similarities, there is of course one crucial difference: the barefoot vet represented the reach of state governance into rural society, but the duck doctor appeared in the wake of the withdrawal and detachment of state veterinary services, as unofficial economies of poultry disease grew amid the vital uncertainties of the livestock revolution. Although I use the term *unofficial* to indicate the duck doctor's unlicensed status, this term provides only a negative description ("nonstate") and does not capture the concrete structures of the unofficial economy. In order to provide a better description, I suggest we might describe the duck doctor as a representative of the growing reach of the livestock-medicine factories (*shouyaochang*) in rural society.

The distributors from the factories often stopped by Hao's shop, checking on his inventory and encouraging him to make further orders. One day when I visited Hao's shop, he received a phone call and, after listening for a minute, used the excuse that he was with a friend to get off the phone. He then turned his phone off. "I have been receiving calls all day!" he complained. They were all from the drug factories. Hao was affiliated with two or three livestock-pharmaceutical companies, he explained. These companies were based in major Chinese cities like Shanghai and Guangzhou. The companies shipped medicines at a regular schedule to a nearby shipping warehouse. The warehouse held the medicines until Hao picked them up. Because he had to pay for the medicines when he picked them up, he tried to leave them at the warehouse for as long as possible. After two weeks, however, they would be sent back to the drug factory. On another occasion, I accompanied Hao as he paid a driver to rush over to the warehouse and pick up a few packages that were about to exceed their two-week window. Today the factory representatives had called him because they believed some of the medicines intended for him had been shipped back, and they were angry. "They are always trying to make me buy more things; it's really irritating," Hao told me.

I once asked the Emergency Center's Cai Haifeng about these duck doctors. Doesn't the duck doctor represent the antithesis of the official veterinarian? I asked him, describing to him diagnostic scenes like the one above. Might the duck doctor suggest an alternative to the problem of professionalization and social distance from the farm that Cai diagnoses in the official veterinarians, including the field epidemiology trainees? Cai grinned and shook his head. These vets who "sell medicines and feeds," he replied, are nothing more than cheats (*pianzi*), quacks, or mountebanks. "They will say anything in order to further their own interests."

However, in the difference between the barefoot vet and the duck doctor we can see something more than the the rise of individualistic entrepreneurship or what Cai saw as self-interested fraud. The increasingly expert, high-quality, technically proficient official veterinarians—a select few replacing the mass of barefoot vets—are increasingly detached from the working landscapes of rural China. The cultivation of official veterinarians is enhancing their professional manner and increasing their expertise in epidemiology or other veterinary sciences. Yet their ability to communicate with, understand, and reach out to rural farmers is not prioritized in the forms of these trainings. The worlds of the official veterinarian and the rural farmer are increasingly distant.

The duck doctor, like the barefoot veterinarian of the past, does share a common world with farmers. Yet unlike the barefoot vet, his livelihood depends on the promotion of pharmaceutical commodities to rural farmers. He is something like an informal "extension agent" for pharmaceutical companies rather than for state agriculture bureaus; he draws on familiarity and shared experience to increase the flow of for-profit pharmaceuticals and to earn fees from each household rather than in order to produce and communicate knowledge about epizootics or eradicate diseases.

Today, veterinary reforms have seemingly split the barefoot veterinarian in two: the office vet and the duck doctor. Cai Haifeng had told me that there are two kinds of veterinarians in China today: those trained in specialized sciences and those who lack knowledge altogether. But the divergence of the office vet and the duck doctor is both more complex and more troubling than this. The office vet has specialized knowledge but lacks familiarity with rural worlds and producers. The duck doctor has familiarity with rural worlds and producers but minimal specialized knowledge. The duck doctor's attachments to all of the beings that people the working landscapes of rural China—duck farmers, feed salesmen, chickens, ducks, and quail—contrasts sharply with the detachment of the official veterinarians, immersed in their labs and laptops. Yet even more troubling is the broader division of technical detachment from human attachment—of specialized knowledge from familiarity with rural production—in two separate forms of veterinary practice. In this book, I have highlighted the transversal pathways, such as those taken by Cai Haifeng in his investigations of Shandong broiler farms, that cross and connect spaces of specialized knowledge and poultry shed life, bridge strata of political governance and working landscapes, thereby producing displacements and forming new assemblages. But as veterinarians are increasingly separated into two types— office vet and duck doctor, detached expertise and enterprising attachment— these transversal pathways are blocked. Scientific expertise and working landscapes separate into layers like oil on water.

CONCLUSION
VANISHING POINT

A point is always a point of origin. But a line of becoming has neither
beginning nor end, departure nor arrival, origin nor destination.
—Gilles Deleuze and Félix Guattari, *A Thousand Plateaus*, 1987

On April 20, 2015, farmworkers at a layer-hen farm in Osceola County, Iowa, began the painstaking cull of four million hens living inside cages at the industrial egg facility. The U.S. Department of Agriculture (USDA) ordered the mandatory depopulation after isolating a new strain of highly pathogenic avian influenza virus known as H5N2 from birds at the farm. Workers pumped a water-based foam into the barns in order to suffocate the birds. Then they covered the birds with litter and other organic material and left them for several days to compost inside their cages. Despite the cull, the outbreak continued to spread. By the end of May, more than two hundred outbreaks had occurred across the Midwest, requiring the destruction of nearly fifty million birds.

When the U.S. Senate convened a hearing on the crisis on July 7, USDA officials blamed the outbreak on the failure to control the virus at its source. But that source, they argued, was not in Osceola County or anywhere else in the United States. That source was in China. Dr. John Clifford, the deputy administrator of the Animal Plant Health Inspection Service, underscored in Senate testimony that the outbreak was "unusual": "for the first time we had a high-path avian influenza virus to cross from Europe and Asia into North America, the first time ever."[1] Drawing on phylogenetic comparisons of influenza virus genomes, Clifford argued that the "parent" of the virus afflicting the Midwest was the highly pathogenic H5N1 strain that first emerged in southern China and Hong Kong in 1997, almost twenty years earlier: "If you all would go back and look at the concerns at the time, it was concern that this would be the next human pandemic. We put some money toward trying to address H5N1 in Asia, but we did not put enough. If the world had put more money toward that effort and addressed these diseases in the animals at the time, we would not have this situation today, because that—what occurred in 1997 was the original finding of that virus in China—has caused this outbreak today."[2]

The hypothesis of the pandemic epicenter I have explored in this book contains a distinctive notion of causality: a point of origin is responsible for events that take place in future times and at distance places. As Clifford phrased it succinctly, if somewhat awkwardly, in his Senate testimony, "the original finding of that virus in China . . . has caused this outbreak today." If the emerging virus was contained at its source, then a future catastrophe would not have occurred. The potential would never be realized. The trouble there would have never gotten here.

In this attribution of cause to origin, many intermediaries are removed: the wild birds that hypothetically brought an H5 virus across the Arctic and into British Columbia; the "backyard" poultry and commercial poultry operations in California or Idaho where viruses apparently reassorted; the hyper-biosecure turkey farms and "layer operations"—some eight or ten stories tall—that fueled and spread the outbreak; and perhaps even the depopulation operations conducted by the Animal and Plant Health Inspection Service of the USDA, which tendered these duties to poorly trained private contractors. By ignoring these intermediaries, the hypothesis of the pandemic epicenter enacts an "erasure of context" in which source and consequence, pathogen emergence and global epidemic, are conflated.[3] Yet at the same time the pandemic epicenter also raises new questions about the context of the source: Why did the virus emerge in China? What is it about "there" that is different from "here"?

Several months after the Senate hearings, a journalist from National Public Radio contacted me regarding my fieldwork at Poyang Lake. NPR's *Marketplace* was "doing a story on Avian Influenza in China," she explained, in response to the 2015 outbreak of avian influenza in the American Midwest. "They think the flu first emerged in China, then made its way to the U.S.," she added. I assumed she would want to interview me for my perspective on the matter, but instead she asked if I had any local contacts in the Poyang Lake region. They were "planning on visiting Poyang Lake later this month" in order to trace the emerging disease to its possible source.

In the broadcast, *Marketplace* journalist Rob Schmitz highlights Poyang Lake's unusual avian ecology, where "more migratory birds spend their winters" than anywhere else in Asia, and local farmers are raising "all sorts of poultry." Because of contacts between wild and domestic birds, Poyang Lake is "a cauldron for brewing new strains of flu." Schmitz visits rice paddies, a "backyard-type" duck farm, and a live-poultry market where "ducks, chickens, pigeons, and other animals lie in cages next to each other." Chinese consumers "prefer to buy live poultry" rather than slaughtered and butchered meat, so Schmitz concludes that China's growing economy will only "increase the risk for the spread of new strains of avian influenza." Poyang Lake was, the title of the NPR broadcast underscored, "the Chinese lake that's ground zero for the bird flu."

As Schmitz's on-the-spot report emphasizes, not just any place can be a pandemic epicenter. According to historian Nick King, the emerging infections worldview typically identifies "the postcolonial economic periphery of 'developing nations' as the source of potential and actual global pandemics."[4] These peripheries are marked according to their exotic ecological character as much as their distant geographic location. The pandemic epicenter is a place of dangerous mixtures, what an older symbolic anthropology might describe as matter out of place.[5] Wild and domestic birds, ducks, chickens, pigeons, "and other animals," the metaphors proliferate: *perfect storm, mixing bowl, mixing vessel, cauldron.* The pandemic epicenter condenses the "changing spaces of globalization and their intrinsic dangers," confusing tradition and modernity, human and nonhuman, nature and culture.[6]

Through the condensation of the pandemic threat into a point of origin, the epicenter hypothesis holds out the promise of containment, or even the prediction and prevention of pandemics "at source." The dangerous mixtures at the pandemic epicenter invoke the redemptive figure of the virus hunter. "There are two ways to discover new pandemics," says biologist Peter Daszak in a call for expanding virus surveillance. "We can get out there and look in wildlife, discover where the viruses are and block transmission. Or we can sit here and wait, and hope we have the right vaccines and drugs."[7]

In this book I have examined what the project for containment at source looks like in practice, beyond the hype or hope for pandemic prediction and prevention. Rather than experts isolating the virus, blocking transmission, and containing the outbreak, I found that the spatial relocation toward the epicenter displaced the model of scientific expertise embodied in the heroic virus hunter. The movement into the epicenter drew scientists and other influenza experts along a line of transformation from laboratory centers in

Geneva, Rome, and Hong Kong into the lakes, rice paddies, and poultry farms of southern China. On the way, the journey confronted scientists with local inhabitants (the epicenter as working landscape) and political boundaries (the epicenter as geopolitical territory), forcing unexpected turns and detours. I showed that these turns caused displacements to the equipment of expertise—including epistemological concepts and models of biosecurity, experimental practices, and norms of scientific exchange. What changed when scientists went into the epicenter?

First, the movement into the epicenter displaced the epistemic conditions of research on pandemic influenza. In his historical reconstruction of twentieth-century disease ecology, Warwick Anderson documents a "minor tradition" of research on the ecology of infectious diseases, carried forward by "holistic" microbiologists such as Macfarlane Burnet and Rene Dubos. However, he notes that Burnet and Dubos adopted ecological frameworks as "metaphoric resources" rather than "analytic tools." Although they frequently framed microbes in terms of host–parasite relations and evolutionary development, they rarely used tools from postwar ecological sciences such as mathematical modeling or systems theory.[8] In part I, I documented how the FAO's projects to stop an emerging influenza pandemic "at source" led to the displacement of ecological metaphors in favor of the modeling analytics of spatial ecology. The colorful invocations of ecology that proliferate in both scientific and journalistic texts on the pandemic epicenter—imagistic accounts of "rice/bird/water/man associations," of multispecies "cauldrons" for brewing viruses, of overcrowded poultry sheds and bloody bird markets—became the object of technical analyses using ecological tools such as spatial modeling, landscape imaging, and species tracking. Although the term *ecology of influenza* persisted, the concept or meaning of the term was displaced by the shift in "epistemic culture"[9] from virology to spatial ecology. At stake is a difference between two epistemologies of ecology, one might say: on the one hand, a microbe-centered ecology that traced connections between the influenza virus and its multiple hosts based on methods of virus sampling, laboratory reassortments, and phylogenetic classification; and on the other, a landscape-scale analysis of the ecological conditions that drive or encourage the emergence of new influenza viruses, from farming practices to wild-bird migrations, and from market trade networks to environmental topography.

On a second plane, the movement into the epicenter displaced the laboratory models of experimental practice that support the heroic figure of the virus hunter. As researchers established experimental systems in places such as China's Poyang Lake, they encountered unexpected objects—such as farmed wild birds—that led them to see natural ecologies and environment as artifacts of human work, or what I have called "working landscapes." I showed how biosecurity models adapted from the U.S. poultry industry failed, leading the FAO to call for "solutions developed locally through direct engagement with stakeholders"; how wild-bird specialists posed as poultry buyers to uncover the extent of wild swan goose farming at Poyang Lake; and how national veterinarians conversed in regional dialect and cultivated friendships with poultry traders to expose the unexpected shipments of spent hens. In each case, knowledge of the "ecology of influenza" was rerouted through the understanding of human practices. "It's interesting how your own mind evolves over time, to see how the way you think about things evolves over time," Marius Gilbert reflected a few years after leaving China. At the epicenter he visited Poyang Lake and was surprised by the "magnitude," the "sheer numbers of poultry," and unusual practices including wild-bird farming. The laboratory's mastery of scale was tempered by the enormous dimensions and intricate recombinations of the virus's actual living environment. He now believed that the ways human beings "work with the environment, for instance through farming," were the most important drivers of disease emergence. At stake was more than the inclusion of humans into a broader cultural ecology of influenza. Instead, I showed how an ethics of expert "detachment"—the laboratory reconfiguration of subjects and objects that, as Knorr-Cetina puts it, "set[s] up a contrast to the surrounding social order"—gave way in favor of efforts to find affinities with surrounding practices that might increase understanding. For a moment, at least, researchers left the lab behind and left the protocols of laboratory practice at the bench.

Scholars have recently pointed to such cracks and fissures in the walls of the laboratory, and the authority of scientific expertise once protected inside those citadels, as an important shift in the relations of science and society.[10] In their influential diagnosis of what they call "Mode 2" science, for instance, Helga Nowotny and her colleagues observed that the sciences, once exclusive and autonomous domains, had now been "superseded by a new paradigm of knowledge production, which was socially distributed, application-oriented, trans-disciplinary, and subject to multiple accountabilities."[11] Previous studies have focused on *political* or *ethical* drivers of this transformation, such as patients' organizations, risk assessments, and ethics regimes.[12] These studies describe new actors entering into the previously circumscribed space of scientific expertise and collaborating in, or contesting, the production of knowledge. The story I have told here is different. In the search for the origin of influenza pandemics, it is the movement of the site or venue of scientific production—

from laboratory centers to the fields of the epicenter—that displaced the scientific expert. Making scientific knowledge at the epicenter did not invite society into the laboratory in response to a politics of participation but rather pushed scientists to exit the laboratory and build new methods for understanding the other creative practices transforming the natural world.

Although this journey into the epicenter is a singular one, the trajectory of centrifugal displacement that I observed is not. Ulrich Beck's call in 1982 for scientific "specializations in the context" as a reflexive response to modernization risks is now almost a commonplace of the contemporary sciences.[13] In fields ranging from genomics, epigenetics, and toxicology to climate change and infectious disease, as well as marine biology and biodiversity conservation, the context or environment is becoming a scientific object in its own right. At the same time as researchers turn surroundings into objects, however, these fields of study are confronting natural environments that are assumed to be in some sense anthropogenic, the product of human works. The contexts spiral outward, along a centrifugal series of displacements, as histories of agricultural and industrial development, the social distribution of technologies, political events such as warfare or revolution, cultures of consumption, and much more become relevant to understanding the scientific object at hand. In these moments that increasingly seem to typify our time in the so-called Anthropocene, scientists are perched at the doors of their laboratories, looking out.[14]

Finally, at a third level, the movement into the epicenter brought global health projects for pandemic preparedness onto China's national territory, producing new reassortments of transnational science and "biosovereignty." In contrast to the figure of the virus hunter, who singlehandedly confronts unruly ecologies and microbial contamination in order to extract pathogens and bring them back to the biosafety of the lab, global health researchers at the influenza epicenter found that access to virus samples required political finesse, compromise, and exchange with China's state agents. Scientists did not simply arrive in the field, take what they need, and leave: as I showed in chapter 5, it was banqueting, offstage agreements, and training exchanges that enabled access through strategies of detour and affinity. Throughout, however, the temptation to stratification remained: the rhetorical separation of ecological region from political territory often served as a resource for denunciation, a technique that could be used to expose a lack of transparency in China's interactions with global health.

I also found that the pursuit of political affinities could unintentionally drive a broader process of stratification. In chapter 6 I described the most ambitious effort of access through affinity: the FAO Emergency Center's field epidemiology training program for state-employed "official" veterinarians (*guanfang shouyi*). By cultivating a modern corps of elite state veterinarians with cosmopolitan outlook and international standard technical knowledge, the FAO hoped to construct a long-term, consistent counterpart for sharing information about disease ecology at the epicenter. In practice, however, training programs for "official veterinarians" encouraged a process of detachment by professionalization, in which cosmopolitan mobility and technical expertise replaced familiarity with the practical and moral worlds of rural China. As a result, precisely as the FAO's training programs enabled access to the geopolitical territories of the epicenter through the intermediary of official veterinarians, these vets were losing their own affinities with the working landscapes of rural China.

I conclude, then, by returning to where I began: the persistent repetition in scientific and popular texts—what, adapting Carlo Caduff's phrase, I might call the serial localization—of China as a pandemic epicenter.[15] China was first identified as a possible source of influenza pandemics in 1957, but the attribution of China as "ground zero" for the flu continues today in scientific publications, newsmagazines, and even feature films. As I noted in the introduction, anthropologists and other scholars have argued that these narratives "too often consist of allegations of blame and assumptions of cultural shortcoming."[16] In a critical discussion of a *Newsweek* article that identifies Guangdong poultry farms as a potential source of influenza pandemics, for instance, Priscilla Wald argues that the magazine "fostered medicalized nativism," a stigmatizing process in which "disease is associated with dangerous practices and behaviors that allegedly mark intrinsic cultural difference."[17] By following scientists along the journey into the epicenter, however, I saw that scientific movement displaced static geographies of blame when it encountered the inhabitants and territories of the epicenter. By the end of the journey, when scientists arrived at the hypothetical location of the epicenter, they found to their surprise that it had already been displaced: the identification of China as a pandemic epicenter was pushed aside by their own trajectory of inquiry.

In chapter 5 I described the Emergency Center's creation of a risk map for H5N1 in China. The map drew on collaborative interdisciplinary research, including Marius Gilbert's spatial modeling, Xiangming Xiao's satellite imaging, and Vincent Martin's negotiated access to China's virus samples, to map the potential distribution of avian influenza disease across China based on a model of ecological and cultural risk factors. I clearly remember the first time I saw these maps. In early spring 2011, I was meeting Martin at his office in the Emergency Center to discuss where I might conduct some field trips to southern China. After listening to my research interests, he suddenly got up from the couch where we were sitting, went to the computer at his desk, and returned a few minutes later with two printouts. The printouts contained several sections of an early version of the H5N1 risk map in close-up detail, focusing on areas around the Yangtze River. The terrain that I looked at was mottled from blue to deep red, a range in hue that I learned stood for variation in the risk of H5N1 infection predicted by the model. Martin began wondering out loud whether studying poultry value chains in the wetlands near urban Wuhan—marked a deep red on the map—might be promising. But as I looked at the images, I was struck by how much this map differed from older concepts and representations of the pandemic epicenter. After the 1957 pandemic, as the WHO and Chinese scientists both traced the origins of the pandemic to China, representations took the form of global or regional maps with a "point of origin" clearly marked in southern China, somewhere in Guizhou Province (see figure I.1). And when Kennedy Shortridge introduced the hypothesis that China was an "influenza epicenter" because of its distinctive agrarian ecosystem, he focused attention and experimental concern on the Pearl River Delta region adjacent to Hong Kong. In contrast, the new risk model described a "wider and slightly displaced epicenter of influenza viruses" in southern China's agricultural landscapes, including previously little-studied areas such as the Yangtze River and Poyang Lake.

In a later conversation, Martin explained to me that with the ecological risk map the "epicenter has become more multifocal by essence": "Even by the end of my stay over there, even through the publications we had, initially it was I think triggered by this idea of epicenter in the south of China, then even from my research, it was showing, not only this southern part of China which was referred to as epicenter by Shortridge a long time ago, but you could see that the south of the Yangtze River was also a very important area for emergence of viruses, so at least the concept of epicenter had evolved even in terms of its geographical coverage."

Furthermore, these ecological maps also made possible an extended spatial representation of risk no longer tied to long-standing ideas about China as an origin point for pandemics. Of particular importance in this regard is the use of new instruments, including satellites, in the mapping of influenza risk. With remote sensing for landscape topography and satellite telemetry for tracking wild birds, the mapping of risk no longer needed to stop at territorial borders. Taking rice-cropping intensity as an exemplary risk factor for HPAI, Julien Cappelle explained to me that remote sensing tools enabled the understanding of the spatial geography of risk to achieve a planetary scale. With remote sensing "you had a powerful tool that can map you the rice-cropping intensity over, well, the world actually," Cappelle noted, "because your satellite images give you data all over the world."

And with this, the concept of a point of origin for pandemic diseases itself finally begins to fade away, surpassed by new ways of engaging and mapping disease emergence that had been developed inside the epicenter. When I spoke with Vincent Martin in 2016, he reflected that when he set up the Emergency Center in Beijing nearly ten years earlier, "we had this concept in HQ or in our animal health program, that we need to tackle the disease at source. I mean that was really something; we had this partnership with OIE, and countering the disease at source was really important: where is the source, where is the origin, and then at some stage when you had such a kind of multifocal emergence of viruses then you were wondering, well, what is a source? Where is a source?"

PANDEMIC PLANET

In 2016 a research group including Marius Gilbert, Scott Newman, and Xiangming Xiao published an article comparing the risk factors that contributed to the spread of H5N1 and what they called "H5Nx" viruses, including the 2015 H5N2 that struck the midwestern United States.[18] The study was based on the methodologies of spatial risk analysis that I discussed in this book. All of these studies have focused on identifying spatial correlations between flu outbreaks and particular risk factors, including poultry density, water presence, and anthropogenic variables (such as population density and distance to roads). However, as the authors explain, there was one crucial difference in the new study. All previous studies were framed within the territorial borders of a particular nation-state, so "studies comparing different sets of factors were never carried out at a global scale." Although techniques such as remote sensing produced data that could be mapped at a planetary scale, in practice the objects of previous studies had still remained confined within national territorial boundaries.

The 2016 paper proposes to produce the "first global suitability map for H5N1 HPAI virus sustained transmission." In breaking through the confines of territorial boundaries, the study displaced the pandemic epicenter onto a planet of continuously variable risk. Red and orange zones of high virulence spill across borders, clustering in regions including central China, northeast Asia, northern Europe, and the midwestern United States and Canada. Such a planetary map marks differences in levels of risk but does not localize a particular point as an epicenter. As the authors explain, these global risk maps identify the suitability of a geographic place for "sustained transmission," not a point of origin. By focusing on "poultry outbreak locations," the study has "limited use in identifying the locations of initial introduction of avian influenza viruses, or places where viruses may undergo more frequent reassortment events leading to the emergence of new viruses." Instead, it provides information about the configurations of cultural practices and multispecies ecologies that may offer affordances, or what they call "suitability," for outbreaks.

During the 1990s ecologists developed habitat suitability models "to predict the likelihood of occurrence of species on the basis of environmental variables."[19] In a few more-recent studies, suitability models diagrammed the future potential for a disease or invasive species to spread into a particular habitat rather than predicting actual occurrence.[20] In this way, global suitability models invert the form of the pandemic epicenter, approaching the potential uncertainty of pandemic preparedness from a new direction. Rather than seeking out the point of first emergence in order to "stop a pandemic there before it gets here," global suitability models diagram the geography of affordance into which that potential new pathogen would arrive. To which landscapes, ecologies, or configurations of human–animal relations would the spread of a disease be possible, if it were to emerge? Would here be more likely than there?

At the beginning of this book I showed how the idea of the influenza pandemic was linked to the moral pursuit of world health and the monitoring of disease across a global scale. In turn, the monitoring of influenza pandemics (in 1957 and 1968) as they spread across the planet enabled a "point of origin" to come into view, although that point remained unfocused and unexplained. As the search for China's pandemic epicenter progressed, the point of origin became defined as an ecological source, both a geographical place and a species reservoir. In chapter 2 I showed how scientific tools such as spatial ecology replaced the metaphoric treatment of the ecology of the epicenter, a process that I described, following Georges Canguilhem, as the displacement of scientific ideology. The more-recent global suitability maps show that the epicenter hypothesis itself was also ideological (in Canguilhem's specific sense) or, perhaps more precisely, an example of what Ludwik Fleck called a "proto-idea." For Fleck, proto-ideas are objects of knowledge and concern that guide the direction of scientific work, shape its development and passage, yet are subsequently displaced by the results of that scientific inquiry. In his study of the invention of the Wasserman blood test for syphilis, Fleck notes that popular ideas associating syphilis with "changes in the blood" or "impure blood," although they are "somewhat hazy" and "emerging from a chaotic mixture of ideas," are "undeniably linked" to and "scientifically embodied" in the subsequent development of the Wasserman reaction.[21] Similarly, a whole range of scientific and popular ideas involved in the accusation of China as an ecological "ground zero" for pandemics motivated and encouraged experimental research projects to map the spatial ecology of influenza risk across China's landscapes. But the very creation of those maps shifted the question away from the localization of a point of origin toward new planetary assessments of suitability, affordance, and vulnerability to viral diseases. In the process these new maps reconfigured the relationships between influenza pandemic, global scale, and scientific inquiry that had framed the initial search for the origins of pandemics.

But the journey I followed here also differs in several respects from the scientific displacements examined in historical epistemology and sociology of science. Whereas Fleck, Canguilhem, and Rheinberger explore displacements within the confines of scientific experiments, I found displacements caused by the spatial movement of experimental systems into the landscapes and territories of the epicenter. "Knowledge, after all, does not rest upon some substratum," Fleck notes. "Only through continual movement and interaction can that drive be maintained which yields ideas and truths."[22] But there are important differences between taking displacement as a metaphor for conceptual movement within experiments, on the one hand, and tracing the consequences of the spatial movement of scientists and their experiments, on the other. One of these differences is that the scientific journey into the epicenter caused displacements to the subjects, as well as the objects, of scientific inquiry.

In their report on the 2016 global suitability models, Gilbert and colleagues offered a provocative claim about what increased suitability for the spread of H5Nx avian influenza viruses. According to their results, H5Nx outbreaks are "best modelled by predictor variables relating to host distribution" rather than "land use or eco-climatic variables." Risk variables such as human population density and density of poultry population, as well as more specific host characteristics such as industrial poultry farming, all offered much better predictions of the distribution of outbreaks than any tested factor related to landscape or climate, such as altitude, wetlands, or temperature variables. In contrast with earlier environmental or ecological theories of disease, such as those applied to vector-borne diseases in twentieth-century tropical medicine, this ecology is not rooted in geographic or climatic regions.[23] Rather, it is modern infrastructures that shape the geographic extension and spatial arrangement of humans and their domestic animals, the most important host populations for the virus. For example, the production and distribution chains of feed manufacturing shape the location and scale of industrial layer farms, not features of the immediate environment such as temperature or humidity. Indeed, feed manufacturing and biosecure poultry sheds are designed to detach poultry from these environmental contexts. As a result, the geography of suitability for H5N1 outbreaks does not map along with climatic or ecological regions such as the tropics, but rather along with the development and distribution of livestock economies.

Here, these global suitability models map a planetary geography of anthropogenic environmental risk that overlaps other recent diagrams of world risk society, threatened Gaia, global climate change, or the Anthropocene age.[24] Scientific accounts of the global environment, Tim Ingold has written, often involve a spatial confusion. The concept of environment, he notes, properly refers to "that which surrounds, and can exist, therefore, only in relation to what is surrounded." But in the concept of a global environment, the environment "is turned in upon itself, so that we who once stood at its centre become first circumferential and are finally expelled from it altogether," observing humanity, life, and the Earth as if from the outside.[25] In Ingold's rendering, as human forms of life and production change the planet, a gap appears between the subject-centered production of working environments and the external point of view from which planetary change is observed. Ingold has described a contrast between "inhabitant knowledges," a "way of knowing [that] is itself a path of movement through the world . . . along a line of travel," and scientific or "occupant knowledge" that builds models out of discontinuous points and is "founded on a categorical distinction between the mechanics of movement and the formation of knowledge, or between locomotion and cognition."[26]

In this book I described how accompanied scientists on their journey into the epicenter in order to show how, in contrast to Ingold's figuration of scientific work, the mechanics of their movement directly contributed to the formation of their knowledge. As they traveled into the epicenter, they made do amid unfamiliar surroundings, interacted with other inhabitants, and encountered the outsides of their experimental knowledge. Ingold's figure of the scientist as "occupant" rather than "inhabitant" is based on the laboratory model of expertise, in which scientific practice is understood as the detachment of facts. For him, scientists and their instruments "drop down" into a "series of fixed locations" to collect "facts" and remove them from their contexts: although a fact may be "discovered among the contents of a site, where it is, or how it came to be there, forms no part of what it is."[27] But when scientists traveled to Poyang Lake, several of their most important findings, those that displaced their experimental systems into new directions, came when they left their laboratory practices behind and sought to understand the sites where they were standing.

To be sure, Gilbert and the others produced the global suitability maps long after they had departed from China and the pandemic epicenter. And at a glance, these maps do appear to invoke a trajectory of stratification—or what Ingold describes as "circumferential" knowledge of the global environment—that transcends the heterogeneity of cultural practices and political borders through detached ecological representations of the globe. As Bruno Latour recently warned regarding the sciences of the Anthropocene more generally, "The danger is always the same: the figure of the Globe authorizes a premature leap to a higher level by confusing the figures of connection with those of totality."[28] But if we treat scientific practice as a line of becoming and displacement rather than laboratory detachment, the expansion toward global scale need not invoke a fantasy of totality. Rather, global maps inscribe new lines of connection across the surface of the Earth. Knowledge of the global environment continues along disparate, forking pathways such as migratory-bird tracking, remote sensing indicators, and statistical population models, and relies on techniques of "planetary-scale fieldwork" to build new connections across scales and ground-truth global representations.[29]

The identification of anthropogenic drivers of environmental risk is not an end point for inquiry any more than the epicenter was a final destination. "We are the most important agent of change," Marius Gilbert reflected when I asked him how the global suitability maps changed his understanding of avian influenza risk. "The first place to look is to look at what people do." The question that remains is the form that such looking will take: the manner in which contemporary global health will place anthropogenic environments under observation and experiment, and the displacements that will result.

FROM GLOBAL TO PLANETARY HEALTH

Today, there is growing agreement that global health must go beyond the reactive containment of emerging outbreaks at pandemic epicenters. Yet there are at least two different visions for what form global health should take in an effort to mitigate the ecology of emerging diseases.

One approach is spearheaded by an initiative known as the Global Virome Project (GVP), made up of participants from organizations including the WHO, the FAO, the U.S. CDC, the non-profit Ecohealth Alliance, virus hunter Nathan Wolfe's private-sector Metabiota, China CDC, and many more. Leading researchers from these organizations launched the GVP in 2018 as a direct response to the failure of outbreak response in global health. "Following each outbreak," write Dennis Carroll and colleagues, "the public health community bemoans a lack of prescience, but after decades of reacting to each event with little focus on mitigation, we remain only marginally protected against the next epidemic."[30] Designed as a program of "large-scale sampling and viral discovery," the GVP plans to isolate previously unknown viruses from a wide range of animal hosts, sequence them in the laboratory, and then use both wet-lab and bioinformatics techniques to assess their risk.[31] Rather than monitor viral outbreaks, the primary objective of the GVP is to construct a "global atlas of most of the planet's naturally occurring potentially zoonotic viruses," many of which have not caused human, livestock or even wildlife outbreaks.[32] By building and researching a large collection of viral agents with zoonotic potential, the GVP aims to achieve the "preemption" of disease emergence.[33] In 2018, some of the first pilot projects began in China, where "scientists will collect virus samples from animals such as rats and bats for study using techniques such as next generation (DNA) sequencing to find whether new viruses exist," reported Gao Fu of the China CDC.[34]

The global virome is global in the same sense as others have described the emergence of a "global biology" in stem-cell research, HIV vaccine development, and the Human Genome Project.[35] On the one hand, geographic locations and animal reservoirs are defined as local sources where viruses are sampled and extracted; on the other, standard protocols for laboratory analysis and data sharing are used to exchange and research virus samples across a transnational domain. A focus of the initiative revolves around developing political protocols and technical infrastructure for data sharing in an effort to create "global ownership," preempting possible claims of viral or genomic sovereignty.[36] Once viruses are assembled into a collection (the virome), machine learning and bioinformatics approaches will predict risk based on virus genotype, enabling the development of countermeasures before any outbreak occurs. Viruses become global to the extent that infrastructures of sampling and sharing enable their "decontextualization and recontextualization, abstractability and movement, across diverse social and cultural situations."[37] In the GVP the global takes the form of a *network*.[38]

By contrast, the assessment of "global suitability" for zoonotic pandemics proposed by the researchers I accompanied in China provides a very different morphological model for global health. It moves beyond the extraction and exchange of viral agents in order to put landscape reconstruction at the core of pandemic preparedness. In a recent paper, Marius Gilbert, Xiao Xiangming, and the FAO's Timothy Robinson reflect on the emergence of avian influenza in China as "a 'One Health/Ecohealth' epitome."[39] Like proponents of the GVP, they begin by remarking on the failure to predict the emergence of several zoonotic diseases over the last few decades, including avian flu, but they offer a very different explanation for this failure. First, they argue, predictions and forecasts are typically based on what is known about a disease and its risk factors "where it circulates" but do not consider the range of factors that "could be important in different areas or under different conditions." In many cases, diseases emerged, or caused unexpectedly severe outbreaks, precisely when the virus suddenly appeared in a new location (for instance, the West Africa Ebola outbreak in 2015). Second, predictions or forecasts are often poorly equipped to monitor "gradual changes in anthropogenic, environmental and wildlife factors" or their interactions and feedback circuits. In their focus on viral discovery and risk assessment, most pandemic-preparedness initiatives ignore how changes in social and ecological environments could transform the scale of threat posed by a particular pathogen.

Gilbert and colleagues identify the industrialization and intensification of livestock farming as a pivotal pressure that drove the emergence and spread of avian influenza viruses in China. But the role of livestock in changing disease landscapes is not "specific to parts of Asia," they add. Growing livestock populations across the world are giving rise to new diseases, from goat expansion in the Netherlands driving the emergence of Q fever to the outbreak of MERS that followed from intensification of camel production in the Middle East. In the face of these landscape pressures, intervention must go beyond early warning or even the preemption of virus emergence. Instead, they argue, global health should work to reduce the suitability of our planetary ecology for the emergence and transmission of zoonotic diseases, including the "deindustrialization" of livestock farming in high-income countries and its "sustainable intensification" in poor countries.[40]

To better understand the implications of this diagnosis for global health, I would like to situate Gilbert and colleagues' proposal in the context of a recent initiative known as *planetary health*.[41] Led by the Rockefeller Foundation in conjunction with the British health journal *The Lancet*, planetary health differs from the many similar-sounding programs—One Health, Ecohealth, ecosystem health—in several important respects. The planetary health approach is based on the idea that human health must be considered in conjunction with the overall health of planetary ecosystems. Arguing that economic development has made the human population "healthier than it has ever been," planetary health reports also note that economic development has created unprecedented threats and risks to planetary ecosystems: pesticide accumulation, depletion of ocean fisheries, and, of course, global climate change. As *Lancet* editors Richard Horton and Selina Lo elaborate, planetary health highlights how human health depends on a "safe operating space" defined by "planetary existence": "if the boundaries of that space are breached, the conditions for our survival will be diminished." But the concept of planetary health also focuses on the *reflexivity* of health threats: "the risks we face lie within ourselves and the societies we have created," Horton and Lo claim.[42]

In these projects, I suggest, global health adopts the form of a sphere rather than a network of standards that enable decontextualization and transfer. Thinking about the globe as a spherical form means considering how the whole Earth, the terrestrial planet, is made into a unified object of knowledge and intervention. By this I do not mean the seventeenth-century image of the Earth as a round globe, a billiard ball in infinite space, but rather the more recent configuration of the Earth as a fragile, bounded sphere within which life is possible.[43] By considering global health within the context of the biosphere, the *interconnections* of human and nonhuman health, the feedback loops between human actions and ecological reactions, are brought to the fore. As Horton and Lo put it, global health action must understand *both* the dependence of human health on its planetary operating space *and* the human, reflexive role in transforming those ecological conditions. In doing so, this spherical form of global health makes visible the planetary-scale risks that we face but also reveals the opportunities to condense rather than amplify feedback loops.

Yet despite the promise of the planetary health vision, it remains largely utopian without concrete infrastructure for both monitoring and reconstructing pathogenic landscapes. In the Global Virome Project, as I showed above, new transnational and transdisciplinary—global—collaborations are built on the foundation of data-sharing protocols and database infrastructures. To achieve the objectives of planetary health, I suggest, requires a different form of global collaboration. The most important divisions are not so much those separating the vertical "silos" between disciplines or the borders between nation-states, but rather the horizontal *stratifications* separating lab sciences from field sciences, bioscience experts from livestock farmers, and humans from nonhumans.

It would be tempting to argue that anthropology, drawing on field research and ever attentive to difference, offers a solution to this problem of stratification.[44] In fact, however, I believe that the figure of the veterinarian offers the best opportunity for building transversal connections between the strata where life is treated as an object of knowledge and those where life is treated as a means of production. Gao Lili, the national veterinarian of the FAO Emergency Center, once offered me an important piece of advice. In a conversation at the Emergency Center, she noted that China's poultry trade was sharply divided between an official sector in which formal contracts between farmers and traders predominated, and a folk (*minjian*) sector where contracts remained unwritten and supported by trust in personal relationships. I interjected that I was considering collaborating with social scientists at one of Beijing's universities to better understand these folk poultry economies. Gao laughed. She said that if I really wanted to know the "details of the field," I should work with local veterinarians, not sociologists. Or even better, find someone who works at a veterinary pharmaceutical or vaccine manufacturer. *They* will know the most farmers.

Yet the veterinarian is not a single figure. In China, for instance, the veterinary sector is increasingly stratified by the separation of official veterinarians, skilled in laboratory and technical knowledges, and the duck doctors who set up their medicine shops at poultry marketplaces. The role of China's livestock industry in the global antimicrobial resistance disaster offers an illuminating recent instance of the dangers of this stratification. On the one hand, official veterinarians in China's Ministry of Agriculture are attempting to enforce new laws prohibiting the use of antimicrobials as growth promoters (including as additives to manufactured feeds). According to recent accounts, however, farmers continue to misuse antimicrobials, including "the ineffective and inappropriate use of the drugs, timing, dosage, and duration, ... [and] the use of counterfeit and substandard products." As Ziping Wu explains, this misuse is primarily because "the advisory role commonly played by veterinarians is increasingly taken over by salespersons of companies that sell their products directly to farmers."[45] Duck doctors—in addition to the obvious profit-seeking motivations—lack the ability to *see* the antibiotic resistance problem, which demands observations at a longitudinal, multispecies, and perhaps even planetary scale. Yet the official veterinarians apparently lack something equally troubling: the ability to understand what farmers are doing and the capacity to advise farmers on better practices.

The collaborations needed for planetary health are not those that enable better data-sharing platforms among disciplines but rather those that increase alliances across heterogeneous domains of knowledge and production. New assemblages, as Deleuze and Guattari proposed, are always initiated by displacements—movements of deterritorialization and decoding—that cross and reconnect different strata.[46] But I would reiterate that there are multiple trajectories of displacement. In an era of planetary health, the next generation of health workers and veterinarians needs something more than the laboratory devices that, like powerful levers, displace natural and social environments, enabling scientists to "raise the world."[47] What they most urgently need is receptiveness to the outside, a capacity to be moved by the world.

On December 31, 2019, China sent a report to the WHO documenting a cluster of cases of severe respiratory pneumonia at a wholesale market in the central city of Wuhan (Hubei Province), and linking the cluster to the emergence of a novel coronavirus. As I write less than two months later, the disease—now officially known as COVID-19—threatens to cause a global pandemic.[1]

As the outbreak grew, international media began to refer to Wuhan as the "epicenter" of the coronavirus outbreak.[2] By the end of January, with the New Year holiday approaching, China responded with intensive measures in an effort to control transmission. Wuhan city was "locked down," with no travelers permitted to leave or enter. Soon after, other secondary cities in Hubei were also locked down. Workers built two makeshift hospitals in a matter of weeks. Cities across China mandated residents to wear surgical face masks in public. Still, the virus spread, reaching countries across Asia, Europe, and North America. Some experts warned that the virus could infect as many as two thirds of the world's population, and could even become a new seasonal scourge like influenza. Yet only a few months before, the virus had never infected a human being.

Once again, global attention focused on the hypothetical source of a dangerous virus: geographically, in China; and ecologically, in animals. The first cluster of cases primarily involved stall owners at the Huanan Seafood Wholesale Market in central Wuhan, who developed severe pneumonia and were warded at hospital intensive care units. Reports soon emerged that the market sold much more than seafood, including a variety of "wild" animals. In samples taken from patients, the Wuhan Institute of Virology isolated a virus with genetic resemblances to SARS-coronavirus and another coronavirus previously isolated from bats, suggesting a wildlife reservoir.[3] Then, researchers from China CDC reported that they isolated COVID-19 virus genes in environmental samples taken from the Huanan market. The virus spilled over into humans from "wild animals at the seafood market," declared Gao Fu, director of the China CDC.[4]

A backlash against the farming, trade, and consumption of wild animals quickly followed in both international and Chinese publics. In the international press, vivid descriptions of "omnivorous" wet markets depicted the sale of civet cats, snakes, and wolf pups in dense crowded stalls covered with water, feathers, and blood.[5] Reiterating the "Orientalist" responses to SARS and avian flu, the emergence of the virus was blamed on "unruly" Chinese consumers who mixed nature and culture in unacceptable ways.[6] But many people in China have also called for a permanent ban on wildlife trading, farming, and consumption in the country. Nineteen scientists from the China Academy of Sciences, including senior virologists from Wuhan, published an open letter on social media calling for the elimination of illegal wildlife trade. "Regulate the illegal wildlife trade at the source, by completely banning the illegal consumption of wild animals," the letter demanded. On January 26, China instituted a temporary national ban on wildlife trading, and in the weeks that followed, police raids led to the arrests of over 700 for violations of the ban. A comprehensive, permanent ban is being considered by the National People's Congress, China's top legislative body.

In ways reminiscent of the responses to pandemic influenza and SARS, scientists transposed laboratory research with viruses onto the landscapes of southern China. Phylogenetic accounts of viral kinship—particularly the resemblance of the COVID-19 virus (isolated from humans) to BatCoV RaTG1 (isolated from horseshoe bats in Yunnan)—were transformed into historical narratives of how the virus emerged from wild animal to human populations. Underlying the dramatic tales of wild animal markets, therefore, is an equally dramatic story about laboratories and international data-sharing. In early January, Chinese labs posted a complete sequence of the COVID-19 virus genome to the EpiFlu database platform managed by the Global Initiative to Share All Influenza Data (GISAID), a database designed to balance wide accessibility with the distribution of credit and other benefits to the originating laboratories without relying on legal claims to viral sovereignty.[7] Teams in Austria, the United States, and Singapore quickly began posting research findings based on bioinformatics modeling that showed the relationships to other viruses, as well identifying possible vaccine and pharmaceutical targets.

In order to develop these comparative inferences, researchers relied on a large archive of coronaviruses previously sampled from bats and other animals in China. Since 2009, the USAID's Predict program and the Global Virome Project have sampled viruses from "bats and rats" in south China. Led by Chinese laboratories, including the Wuhan Institute of Virology, these viral discovery programs collected at least fifteen novel coronaviruses from bats, including eleven isolated from a single bat cave in south China's Yunnan Province.[8] More broadly, the sequencing skills and capacity demonstrated in the response to the outbreak built on China's enormous investment in laboratory sciences—including the construction of a biosafety level 4 laboratory, capable of working with the most dangerous pathogens, at Wuhan. Finally, international sharing of the genome sequences drew on China's previous negotiations with the GISAID database, and contrasted with the "competitive global coordination" that characterized the sequencing race during SARS.[9] The response to the coronavirus outbreak demonstrated the success of what I called a *network* model of global health, in which large-scale collection of viruses and data-sharing across borders enable rapid response to emerging viruses—if not, in this case, their prediction and prevention.

Yet just as the outbreak demonstrated the success of the global health programs devoted to collecting and sequencing viruses, it also exposed the relative paucity of cultural and ecological knowledge of the context of viral emergence. Detailed models of the molecular structure of the COVID-19 coronavirus have been published on the GISAID website; thickly branching phylogenetic trees showing the relatedness of hundreds of virus samples sequenced since the onset of the epidemic are widely circulated; but a basic fact like the number and variety of animal species sold at the Huanan wholesale market is still unknown. Crucial questions that would enable understanding how a virus that previously infected bats "spilled over" into other animal and then human populations are hardly asked, let alone answered: where are wild animal farms located and distributed across China's rural landscapes?; where do farms overlap with the ranges of wild bats?; what new or changing farming practices may have increased contacts across the domestic–wild interface?; which formal and informal trading networks moved animals from farms to Wuhan? As Marius Gilbert and colleagues wrote, viral discovery leads us to identify viruses with the location of their host reservoirs—in this case, bat caves in south China—but we ignore the "gradual changes in anthropogenic environmental and wildlife factors" that could lead one of these viruses to emerge rather than others, often in locations far from their original reservoir.[10] We need better diagnostics of the ecological suitability for spillover, maps of the highways and bridges of viral traffic between wild animals and humans, and not only archives of viruses collected from wildlife reservoirs.

The widespread calls to permanently ban all wildlife trade, including wildlife farming, may be useful in the immediate context of the outbreak, particularly since the relevant host species is still unknown. Ironically, however, calls for a permanent ban also reflect our limited cultural and ecological knowledge of human–wildlife relations. A comprehensive ban would encompass an enormous variety of species farmed and traded under China's "special-type husbandry" (*tezhong yangzhi*) rules, including everything from frogs, turkeys, and quails to civet cats, bamboo rats, tigers, and bears. There is no doubt that some of these species do pose zoonotic disease risks, and a few have even been linked to the emergence of viral diseases (such as civet cats, suspected to play a role in the SARS outbreak). But others (frogs, turkeys) are not considered significant health threats at all.

Banning wildlife markets is also likely to drive them underground, leading to other unexpected risks. In 2013, several cities in eastern China closed live poultry markets after the emergence of a novel strain of avian influenza A(H7N9) in humans. Although the closure of the markets did reduce human cases in the cities in the short term, an FAO study showed that it also led to the expansion of the outbreak by disrupting the typical poultry trade flow, causing the emergence of new trade network patterns and spreading the virus to previously uninfected areas.[11] Two years after live poultry markets were banned in Guangzhou, underground black markets were thriving.[12]

Finally, a comprehensive ban on wildlife economies could further expand the stratification of knowledge and production that I diagnosed in this book, making the forms and patterns of wild animal farming and consumption even less transparent to veterinary governance than they are now. Cultural practices and ecological changes need to be spotlighted with as much care as viruses are examined under the microscope, and this will require ground-up collaboration with farmers, traders, and consumers rather than a social process of scapegoating and exclusion. If there is any hope in predicting and preventing the next emerging viral pathogen, it will rely on inventing new instruments of ecological and social research that are capable of tracing the links between viruses and their environments, reimagining the classificatory oppositions of wild and domestic, and, most difficult of all, working across the stratigraphic divide that separates virological and veterinary experts from rural farm producers.

INTRODUCTION

1 Although this book focuses on pandemic influenza, contemporary global health approaches to the challenge of emerging diseases, such as the U.S. Agency for International Development's Emerging Pandemic Threats Program, invoke a similar logic.

2 Carlo Caduff, *The Pandemic Perhaps: Dramatic Events in a Public Culture of Danger* (Berkeley: University of California Press, 2015); Andrew Lakoff, *Unprepared: Global Health in a Time of Emergency* (Oakland: University of California Press, 2018); Katherine Mason, *Infectious Change: Reinventing Chinese Public Health after an Epidemic* (Palo Alto, CA: Stanford University Press, 2016); Vincanne Adams, Michelle Murphy, and Adele E. Clarke, "Anticipation: Technoscience, Life, Affect, Temporality," *Subjectivity* 28, no. 1 (2009): 246–65.

3 Lakoff, *Unprepared*, 73.

4 Malik Peiris, personal communication.

5 Pete Davies, "The Plague in Waiting," *Guardian*, August 7, 1999.

6 J. C. de Jong et al., "A Pandemic Warning?," *Nature* 389, no. 6651 (October 1997): 554, doi.org/10.1038/39218.

7 FAO/WHO/OIE, "Unprecedented Spread of Avian Influenza Requires Broad Collaboration," January 27, 2004, www.fao.org/newsroom/en/news/2004/35988/index.html. Celia Lowe points out that constructing avian influenza as a global-scale threat "raised the stakes of the problem and made targeted groups of people responsible for the health and well-being of other far-flung humans on the planet." Lowe, "Viral Clouds: Becoming H5N1 in Indonesia," *Cultural Anthropology* 25, no. 4 (2010): 625–49.

8 The title "One World, One Health" was first introduced by the Wildlife Conservation Society (WCS) as part of its "Manhattan principles" in 2004. After the WCS trademarked the phrase, most other organizations referred only to the short form, "One Health." See Yu-Ju Chien, "How Did International Agencies Perceive the Avian Influenza Problem? The Adoption and Manufacture of the 'One World, One Health' Framework," *Sociology of Health and Illness* 35, no. 2 (February 2013): 213–26, doi.org/10.1111/j.1467-9566.2012.01534.x; Abigail Woods et al., *Animals and the Shaping of Modern Medicine: One Health and Its Histories* (London: Palgrave Macmillan, 2017); Susan Craddock and Steve Hinchliffe, "One World, One Health? Social Science Engagements with the One Health Agenda," *Social Science and Medicine* 129 (2015): 2, doi.org/10.1016/j.socscimed.2014.11.016.

9 FAO, "Avian Influenza: Stop the Risk for Humans and Animals at Source," donor appeal, circa 2005, www.fao.org/avianflu/documents/donor.pdf.

10 Phil Harris, "Avian Influenza: An Animal Health Issue," August 2006, www.fao.org /avianflu/en/issue.html.

11 Quoted in Ian Scoones and Paul Forster, "The International Response to Highly Pathogenic Avian Influenza: Science, Policy and Practice" (STEPS Working Paper 10, STEPS Centre, Brighton, 2008, 46).

12 David P. Fidler, "Germs, Governance, and Global Public Health in the Wake of SARS," *Journal of Clinical Investigation* 113, no. 6 (March 15, 2004): 799, doi.org/10.1172 /JCI200421328. See also the discussion in Lakoff, *Unprepared.*

13 David P. Fidler, "SARS: Political Pathology of the First Post-Westphalian Pathogen," *Journal of Law, Medicine and Ethics* 31, no. 4 (2003).

14 "Draconian techniques" from Arthur Kleinman and James L. Watson, "Introduction: SARS in Social and Historical Context," in SARS *in China: Prelude to Pandemic?*, ed. Kleinman and Watson (Stanford, CA: Stanford University Press, 2006), 4; "global hero" from Joan Kaufman, "SARS and China's Health-Care Response: Better to Be Both Red and Expert!," in SARS *in China: Prelude to Pandemic?*, ed. Arthur Kleinman and James L. Watson (Stanford, CA: Stanford University Press, 2006).

15 Mason, *Infectious Change.*

16 "China Is Committed to the Fight against H5N1 Avian Influenza, but Challenges Remain; Opening Remarks by Dr. Shigeru Omi, WHO Regional Director for the Western Pacific Region, News Conference on China's Response to Avian Influenza," PR *Newswire*, December 23, 2005, www.prnewswire.com/news-releases/china -is-committed-to-the-fight-against-h5n1-avian-influenza-but-challenges-remain -55655887.html.

17 Aihwa Ong, "Introduction: An Analysis of Biotechnology and Ethics at Multiple Scales," in *Asian Biotech: Ethics and Communities of Fate*, ed. Aihwa Ong and Nancy N. Chen (Durham, NC: Duke University Press, 2010); Amy Hinterberger and Natalie Porter, "Genomic and Viral Sovereignty: Tethering the Materials of Global Biomedicine," *Public Culture* 27, no. 2–76 (May 1, 2015): 361–86, doi.org/10.1215/08992363 -2841904.

18 Didier Fassin and Mariella Pandolfi, *Contemporary States of Emergency: The Politics of Military and Humanitarian Inventions* (Cambridge, MA: Zone, 2013); Peter Redfield, *Life in Crisis: The Ethical Journey of Doctors without Borders* (Berkeley: University of California Press, 2014).

19 Michael M. J. Fischer, *Anthropological Futures* (Durham, NC: Duke University Press, 2009); for "differentiated sovereignty," see Aihwa Ong, *Flexible Citizenship: The Cultural Logics of Transnationality* (Durham, NC: Duke University Press, 1999), 232.

20 For an analysis of the literary form of "outbreak narratives," see Priscilla Wald, *Contagious: Cultures, Carriers, and the Outbreak Narrative.* For a small sample of tales about virus hunting, some autobiographical, see Nathan Wolfe, *The Viral Storm*; David Quammen, *Spillover*; and Richard Preston, *The Hot Zone.* For "fringes of the nonhuman world," see Maryn McKenna, "The Race to Find the Next Pandemic— Before It Finds Us," *Wired*, April 12, 2018, https://www.wired.com/story/the-race-to -find-the-next-pandemic-before-it-finds-us.

21 On the dominance of microbe-centered research over the "minor tradition" of disease ecology in the twentieth century, see Warwick Anderson, "Natural Histories of Infectious Disease."

22 Paul Rabinow introduced the concept of "zones of virulence" in early formulations of his work on global biopolitics of security. Personal communication.

23 Paraphrasing Hannah Landecker. "Biology of the in-between" and "relational biology" are from Landecker, "Food as Exposure: Nutritional Epigenetics and the New Metabolism," *BioSocieties* 6, no. 2 (2011): 167–94.

24 Georges Canguilhem, analyzing the biology of von Uexkull and Kurt Goldstein, writes that "to study a living thing in experimentally constructed conditions is to make a milieu for it, to impose a milieu upon it; yet, it is characteristic of the living that it makes its milieu for itself, that it composes its milieu." Canguilhem, "The Living and Its Milieu," in *Science, Reason, Modernity: Readings for an Anthropology of the Contemporary*, ed. Anthony Stavrianakis, Gaymon Bennett, and Lyle Fearnley (New York: Fordham University Press, 2015), 181. Canguilhem describes the need for a "biological sense" that attends to the distinctive way that the living agent composes and engages its environment.

25 Notes from 2nd Annual International Workshop on Community-Based Data Synthesis, Analysis and Modeling of Highly Pathogenic Avian Influenza in Asia, Beijing, 2010.

26 Nanjing Zeng et al., "New Bird Records and Bird Diversity of Poyang Lake National Nature Reserve, Jiangxi Province, China," *Pakistan Journal of Zoology* 50, no. 4 (2018): 1199–600.

27 On "working landscape," see Robert Kohler, "Prospects," in *Knowing Global Environments: New Historical Perspectives on the Field Sciences*, ed. Jeremy Vetter (New Brunswick, NJ: Rutgers University Press, 2010). See also Stephen J. Lansing's similar "engineered landscape," in Lansing, *Priests and Programmers: Technologies of Power in the Engineered Landscape of Bali* (Princeton, NJ: Princeton University Press, 2009), 9. On nature as artifact, see Paul Rabinow, "Artificiality and Enlightenment," in Rabinow, *Essays on the Anthropology of Reason* (Princeton, NJ: Princeton University Press, 1996), 91–111.

28 Xiuqing Zou, *Poyanghuqu nongye: ziran ziyuan liyong yanbian jizhi yanjiu* (Nanchang: Jiangxi Renmin Chubanshe, 2008), 117; see also the discussion in Miriam Gross, *Farewell to the God of Plague: Chairman Mao's Campaign to Deworm China* (Oakland: University of California Press, 2016).

29 Christopher L. Delgado, *Livestock to 2020: The Next Food Revolution* (Rome: Food and Agriculture Organization of the United Nations, 1999); David Goodman, Bernardo Sorj, and John Wilkinson, *From Farming to Biotechnology: A Theory of Agro-industrial Development* (New York: Basil Blackwell, 1987).

30 Thus, some collectives focused on raising ducks at a fairly large scale, including in agricultural experiments with biological control of pests. But these were hardly typical. See Sigrid Schmalzer, *Green Revolution, Red Revolution: Scientific Farming in Socialist China* (Chicago: University of Chicago Press, 2016).

31 Jonathan Unger, *The Transformation of Rural China* (Armonk, NY: M. E. Sharpe, 2002).

32 Yougui Zheng and Li Chenggui, eds., 《一号文件与中国农村改革》, 合肥市: 安徽人民出版社, 2008.

33 Statistics are from FAOSTAT. Thomas P. Van Boeckel et al., "Modelling the Distribution of Domestic Ducks in Monsoon Asia," *Agriculture, Ecosystems and Environment* 141, nos. 3–4 (May 1, 2011): 373–80, doi.org/10.1016/j.agee.2011.04.013.

34 Xiuqing Zou, *Poyanghuqu nongye*, 136.

35 Julien Cappelle et al., "Risks of Avian Influenza Transmission in Areas of Intensive Free-Ranging Duck Production with Wild Waterfowl," *EcoHealth* 11 (2014): 109–19, doi:10.1007/s10393-014-0914-2.

36 Quoted by Rob Schmitz, "The Chinese Lake That's Ground Zero for the Bird Flu," *Marketplace: National Public Radio*, March 30, 2016, www.marketplace.org/2016/03/03/world/chinese-lake-has-become-ground-zero-bird-flu.

37 The contemporary dynamics of China's rural working landscapes reflect the conditions of unintended consequences and reflexivity that Ulrich Beck calls "risk society," Kim Fortun calls "late industrialism," and Hannah Landecker calls the "biology of history." Ulrich Beck, *Risk Society: Towards a New Modernity* (London: SAGE, 1992); Kim Fortun, "Ethnography in Late Industrialism," *Cultural Anthropology* 27, no. 3 (2012): 446–64; Hannah Landecker, "Antibiotic Resistance and the Biology of History," *Body and Society* 22, no. 4 (2016): 19–52.

38 Hans-Jörg Rheinberger, *Toward a History of Epistemic Things: Synthesizing Proteins in the Test Tube* (Stanford, CA: Stanford University Press, 1997).

39 Bruno Latour, "On Technical Mediation—Philosophy, Sociology, Geneaology," *Common Knowledge* 3, no. 2 (1994): 32.

40 Bruno Latour, "Give Me a Laboratory and I Will Raise the World," in *Science Observed: Perspectives on the Social Study of Science*, ed. Karin D. Knorr-Cetina and Michael Mulkay (London: SAGE, 1983), 151–52; see also Bruno Latour, *The Pasteurization of France* (Cambridge, MA: Harvard University Press, 1988).

41 Bruno Latour, *Science in Action: How to Follow Scientists and Engineers through Society* (Cambridge, MA: Harvard University Press, 1987).

42 Latour, "Give Me a Laboratory," 153.

43 Other notable laboratory ethnographies include Karin Knorr-Cetina, *The Manufacture of Knowledge: An Essay on the Constructivist and Contextual Nature of Science*, ed. Sheila Jasanoff, Gerald E. Markle, James C. Peterson, and Trevor Pinch (Oxford: Pergamon, 1981); Sharon Traweek, *Beamtimes and Lifetimes: The World of High Energy Physicists* (Cambridge, MA: Harvard University Press, 1992); Paul Rabinow, *Making PCR: A Story of Biotechnology* (Chicago: University of Chicago Press, 1996). For review, see Karin Knorr-Cetina, "Laboratory Studies: The Cultural Approach to the Study of Science," in *Handbook of Science and Technology Studies* (Thousand Oaks, CA: SAGE, 1995).

44 Karin Knorr-Cetina, "The Couch, the Cathedral, and the Laboratory: On the Relationship between Experiment and Laboratory in Science," in *Science as Practice and Culture*, ed. Andrew Pickering (Chicago: University of Chicago Press, 1992), 114–16; Hannah Landecker, "The Matter of Practice in the Historiography of the Experimental Life Sciences," in *Handbook of the Historiography of Biology*, ed.

Michael Dietrich, Mark Borrello, and Oren Harman (Cham, Switzerland: Springer International, 2018), 1:2.

45 Knorr-Cetina, "The Couch, the Cathedral, and the Laboratory," 134.

46 Latour, "Give Me a Laboratory," 166.

47 Bruno Latour, *Pandora's Hope: Essays on the Reality of Science Studies* (Cambridge, MA: Harvard University Press, 1999), 43.

48 Henrika Kuklick and Robert E. Kohler, "Science in the Field," *Osiris* 11 (1996): 1–265.

49 Cori Hayden, *When Nature Goes Public: The Making and Unmaking of Bioprospecting in Mexico* (Princeton, NJ: Princeton University Press, 2003), 186.

50 Knorr-Cetina, "The Couch, the Cathedral, and the Laboratory."

51 Latour, "Give Me a Laboratory," 147.

52 Wen-Yuan Lin has recently articulated an alternate concept of displacement to explain how Taiwanese dialysis patients seek nonbiomedical treatments, such as Chinese medicine, in conjunction with clinic-based dialysis. He argues that patients do not seek to confront hegemonic biomedical frameworks through "trials of strength" but rather to "displace" them by avoiding confrontation, speech, and representation. Lin, "Displacement Agency: The Enactment of Patients' Agency in and beyond Haemodialysis Practices," *Science, Technology and Human Values* 38, no. 3 (2012): 421–43.

53 Joan Fujimura, "Crafting Science: Standardized Packages, Boundary Objects, and 'Translation,'" in *Science as Practice and Culture*, ed. Andrew Pickering (Chicago: University of Chicago Press, 2010), 171.

54 Concepts of strata, stratification, and lines of becoming or transformation are adopted, though rather loosely or disrespectfully, from Gilles Deleuze and Félix Guattari, *A Thousand Plateaus*, trans. Brian Massumi (Minneapolis: University of Minnesota Press, 1987), especially "The Geology of Morals."

55 Hans-Jörg Rheinberger, *An Epistemology of the Concrete: Twentieth-Century Histories of Life* (Durham, NC: Duke University Press, 2010), 8.

56 George E. Marcus, "Ethnography in/of the World System: The Emergence of Multi-Sited Ethnography," *Annual Review of Anthropology* 24 (1995): 95–117; Kim Fortun, "Ethnography in/of/as Open Systems," *Reviews in Anthropology* 32, no. 2 (January 2003): 171–90, doi.org/10.1080/00988150390197695.

57 Paul Rabinow, *French DNA: Trouble in Purgatory* (Chicago: University of Chicago Press, 1999).

58 Caduff, *The Pandemic Perhaps*.

59 A. M. Kleinman et al., "Asian Flus in Ethnographic and Political Context: A Biosocial Approach," *Anthropology and Medicine* 15, no. 1 (2008): 1, dx.doi.org/10.1080/13648470801918968.

CHAPTER ONE: THE ORIGINS OF PANDEMICS

1 Mark Harrison, "Pandemics," in *The Routledge History of Disease*, ed. Mark Jackson (London: Routledge, 2017), 129–46.

2 Mark Honigsbaum, *A History of the Great Influenza Pandemics: Death, Panic and Hysteria, 1830-1920* (London: I. B. Tauris, 2014).

3 Ditmar Finkler, "Influenza," in *Twentieth Century Practice: An International Encyclopedia of Modern Medical Science by Leading Authorities of Europe and America*, ed. Thomas L. Stedman (New York: Publisher's Printing, 1898), 15:1.

4 Peter Sloterdijk, *Spheres,* vol. 2, *Globes: Macrospherology* (New York: Semiotext(e), 2011).

5 Peter Sloterdijk, *In the World Interior of Capital: Towards a Philosophical Theory of Globalization*, trans. Wieland Hoban (Cambridge: Polity, 2015), 22.

6 Paul N. Edwards, *A Vast Machine: Computer Models, Climate Data, and the Politics of Global Warming* (Cambridge, MA: MIT Press, 2010).

7 Emmanual Le Roy Ladurie, "A Concept: The Unification of the Globe by Disease," in Le Roy Ladurie, *The Mind and Method of the Historian* (Chicago: University of Chicago Press, 1981).

8 Andrew Cunningham, "Transforming Plague: The Laboratory and the Identity of Infectious Disease," in *The Laboratory Revolution in Medicine*, ed. Andrew Cunningham and Perry Williams (Cambridge: Cambridge University Press, 2002).

9 On the multiplicity of "germ theories of disease," see Michael Worboys, *Spreading Germs: Disease Theories and Medical Practice in Britain, 1865-1900* (Cambridge: Cambridge University Press, 2000).

10 Andrew J. Mendelsohn, "From Eradication to Equilibrium: How Epidemics Became Complex after World War I," in *Greater Than the Parts: Holism in Biomedicine, 1920-1950*, ed. C. Lawrence and G. Weisz (Oxford: Oxford University Press, 1998), 303-31.

11 Michael Bresalier, "Fighting Flu: Military Pathology, Vaccines, and the Conflicted Identity of the 1918-19 Pandemic in Britain," *Journal of the History of Medicine and Allied Sciences* 68, no. 1 (January 1, 2013): 124, doi.org/10.1093/jhmas/jrr041.

12 Michael Bresalier, "Uses of a Pandemic: Forging the Identities of Influenza and Virus Research in Interwar Britain," *Social History of Medicine* 25, no. 2 (2011): 401, doi.org/10.1093/shm/hkr162.

13 Michael Bresalier stresses the importance of animal models as technologies for making filter-passing viruses visible. The background to the NIMR experimental system was two research programs that successfully isolated viral agents responsible for animal diseases: Laidlaw's work on dog distemper and Richard Shope's isolation of a virus responsible for "swine flu." See also Carlo Caduff, *The Pandemic Perhaps: Dramatic Events in a Public Culture of Danger* (Berkeley: University of California Press, 2015).

14 Michael Bresalier, "Neutralizing Flu: 'Immunological Devices' and the Making of a Virus Disease," in *Crafting Immunity: Working Histories of Clinical Immunology*, ed. Pauline Mazumdar, Kenton Kroker, and Jennifer Keelan (London: Ashgate, 2008), 135.

15 Bresalier, "Neutralizing Flu," 135.

16 Alfred W. Crosby, *America's Forgotten Pandemic: The Influenza of 1918* (Cambridge: Cambridge University Press, 2003).

17 John M. Eyler, "De Kruif's Boast: Vaccine Trials and the Construction of a Virus," *Bulletin of the History of Medicine* 80, no. 3 (2006): 433–36, doi.org/10.1353/bhm.2006 .0092.

18 The effect is caused by the hemagglutinin protein on the outer shell of the influenza virus, although Hirst was unaware of this at the time.

19 In this section I rely closely on the accounts of Hirst's research presented by Carlo Caduff, *The Pandemic Perhaps*, and Michael Bresalier, "Neutralizing Flu."

20 George Dehner, *Influenza: A Century of Science and Public Health Response* (Pittsburgh: University of Pittsburgh Press, 2012).

21 For a time, some American scientists insisted that the variation of influenza was constrained within a limited set of three or more types. However, the growing consensus suggested that the variation was very great or even infinite and that public health efforts to develop a vaccine needed to be closely coupled with surveillance and detection of novel influenza variants. See Eyler, "De Kruif's Boast," 433–36.

22 W. I. B. Beveridge, "Where Did Red Flu Come From?," *New Scientist* (March 23, 1978): 790.

23 Michael Bresalier, "Sharing Viruses and Vaccines: Economies of Exchange in Global Influenza Control, 1947–1957," unpublished paper presented at Universite de Strasbourg, December 6–7, 2012.

24 Kelley Lee, *The World Health Organization (WHO)* (London: Routledge, 2009), 12. On shifting priorities in the WHO at the time between "social medicine" and "technical approaches to health and development," see Randall M. Packard, *A History of Global Health: Interventions into the Lives of Other Peoples* (Baltimore: Johns Hopkins University Press, 2016).

25 WHO, Official Records of the World Health Organization, "Minutes of the Fourth Session of the Interim Commission," 1947, 194. See also A. M. Payne, "The Influenza Programme of WHO," *Bulletin of the World Health Organization* 8 (1953): 755–74.

26 M. M. Kaplan, "The Role of the World Health Organization in the Study of Influenza," *Philosophical Transactions of the Royal Society of London. Series B, Biological Sciences* 288, no. 1029 (1980): 418.

27 M. E. Kitler, P. Gavinio, and D. Lavanchy, "Influenza and the Work of the World Health Organization," *Vaccine* 20, Suppl. 2 (2002): S5–S15.

28 WHO, Official Records of the World Health Organization, "Minutes of the Fourth Session," 194.

29 Michael Bresalier, "Sharing Viruses and Vaccines: Economies of Exchange in Global Influenza Control, 1947–1957," unpublished paper presented at Universite de Strasbourg, December 6–7, 2012.

30 Bresalier, "Sharing Viruses and Vaccines," 16. The Andrewes quotation is from the 1947 Copenhagen meeting.

31 Payne, "The Influenza Programme of WHO," 756.

32 Dehner, *Influenza: A Century of Science*; Frédéric Vagneron, "Surveiller et s'unir? The Role of WHO in the First International Mobilizations around an Animal Reservoir of Influenza," *Revue d'anthropologie des connaissances* 9, no. 2 (2015): 139–62.

33 Andrewes, speaking in 1955, quoted in Bresalier, "Sharing Viruses and Vaccines." In the early years of the WIP there was uncertainty about whether influenza spread from country to country or whether favorable conditions led to an "activation" of the virus. Drawing from surveillance conducted by the WIP in 1948–49, Chinese virologist Chi-Ming Chu, C. H. Andrewes, and other WHO scientists demonstrated that influenza indeed spread from one place to another. C. M. Chu, C. H. Andrewes, and A. W. Gledhil, "Influenza in 1948-1949," *Bulletin of the World Health Organization* 3 (1950): 208–9. This was written when Chu Chi-ming (Zhu Jiming) was working with Andrewes in London before returning to China after the Communist Revolution.

34 WHO Archives, ARCO10-3, Centralized Files, 3rd Generation, Sub-fonds 3, Box 12-418-412, Hale to Payne, May 5, 1957. A few days earlier, the U.S. Influenza Centre had also received a report from Hong Kong stating that an epidemic of respiratory disease occurring primarily among refugees from China had been "confirmed as influenza." World Health Organization, *Weekly Epidemiological Record* 19 (May 10, 1957): 241. The earliest indication of an outbreak actually came from a news report in the *South China Morning Post* in mid-April. The paper also gave hints of the origin of the outbreak in mainland China. The article, dated April 17, 1957, reported that "arrivals in Hongkong yesterday from Central China said that influenza had reached epidemic proportions in Shanghai, Nanking, Wuhan as well as Peking." However, the WHO was not aware of this report until after the other, official telegrams came in. According to Dehner, American researchers first learned of the outbreak through the coincidental reading of a "filler" article in a daily newspaper. See Dehner, *A Century of Science*.

35 K. A. Lim et al., "Influenza Outbreak in Singapore," *Lancet* 273 (1957): 791–96.

36 WHO Archives, ARCO10-3, Centralized Files, 3rd Generation, Sub-fonds 3, Box 12-418-412, Memo to All Influenza Centres, May 24, 1957. See similar account in World Health Organization, *Weekly Epidemiological Record* 22 (May 29, 1957): 277.

37 See World Health Organization, *Weekly Epidemiological Record* 23-52 (1957) and 1–13 (1958).

38 WHO, "Influenza: Introduction." *Bulletin of the World Health Organization* 20, no. 2–3 (1959): 183–85.

39 See Amy L. S. Staples, *The Birth of Development: How the World Bank, Food and Agriculture Organization, and World Health Organization Changed the World, 1945-1965* (Kent, OH: Kent State University Press, 2006), 150. On May 5, 1950, the Republic of China telegraphed the WHO from Taiwan to state that it would withdraw from the WHO because of its inability to meet financial obligations. A week later, the PRC sent a telegram to the WHO claiming to be "the only legal government representing the Chinese people" (cited in Staples, *Birth of Development*, 150). But the World Health Assembly refused to recognize the ROC withdrawal and refused to respond to the PRC's claims. Three years later, the ROC was brought back to the WHO. It would not be until 1972 that China (PRC) rejoined the WHO (and Taiwan/ROC removed), part of which I discuss in subsequent sections of this chapter.

40 The Soviet Union boycotted the UN over the representation of China issue between January 1950 and August 1950.

41 A. M. Payne, "Some Aspects of the Epidemiology of the 1957 Influenza Pandemic," *Proceedings of the Royal Society of Medicine* 51, no. 1009 (1958): 29.

42 "Zhu Jiming," *Biographies of China Science and Technology Experts—Medical Edition, Fundamental Medicine*. China Reference Book Web Publishing Repository, 《朱既明—中国科学技术专家传略·医学编 基础医学卷 一—中国工具书网络出版传略》.

43 Jiming Zhu, Xiao Jun, and Zhengzhang Hao (朱既明, 萧俊, 郝成章), 1957 年流行性感冒流行的病毒类型和性状 I. 长春分离的病毒, 《科学通报》June 30, 1957, 374.

44 Chi-Ming Chu, "The Etiology and Epidemiology of Influenza: An Analysis of the 1957 Epidemic," *Journal of Hygiene, Epidemiology, Microbiology, and Immunology* 2, no. 1 (1958): 1–8. A note on the last page indicates the article was received by the journal on December 2, 1957.

45 Chi-Ming Chu, "The Etiology and Epidemiology of Influenza," 3.

46 Chi-Ming Chu, "The Etiology and Epidemiology of Influenza," 6.

47 Press release cited in "The 1957 Influenza Epidemic." *Indian Journal of Pediatrics,* 24, no. 117 (1957): 354–55.

48 Payne, "Some Aspects," 1009.

49 Cited in Bresalier, "Neutralizing Flu," 120.

50 François Delaporte, *Chagas Disease: History of a Continent's Scourge* (New York: Fordham University Press, 2012), 133.

51 M. M. Kaplan and W. I. B. Beveridge, "WHO Coordinated Research on the Role of Animals in Influenza Epidemiology," *Bulletin of* WHO 47, no. 4 (1972): 439, 442.

52 "非常显着的变异." Jiming Zhu, Xiao Jun, and Zhengzhang Hao, "1957 年流行性感冒流行的病毒类型和性状," 374.

53 Chu Chi-Ming, "The Etiology and Epidemiology of Influenza," 3.

54 WHO Archives, ARCO10-3, Centralized Files, 3rd Generation, Sub-fonds 3, Box Z2/181/44, Kaplan to Laver, May 14, 1970.

55 M. M. Kaplan and A. M.-M. Payne, "Serological Survey in Animals for Type A Influenza in Relation to the 1957 Pandemic," *Bulletin of the World Health Organization* 20 (1959): 466.

56 Kaplan and Payne, "Serological Survey in Animals for Type A Influenza in Relation to the 1957 Pandemic," 482.

57 W. Charles Cockburn, P. J. Delon, and W. Ferreira, "Origin and Progress of the 1968–69 Hong Kong Influenza Epidemic," *Bulletin of the World Health Organization* 41, nos. 3-4-5 (1969): 345–46.

58 Cockburn, Delon, and Ferreira, "Origin and Progress," 346.

59 Vagneron, "Surveiller et s'unir?," 139–61.

60 M. M. Kaplan, "Relationships between Animal and Human Influenza," *Bulletin of the World Health Organization* 41, nos. 3-4-5 (1969): 2393b.

61 C. H. Stuart-Harris, in World Health Organization, "General Discussion—Session II," *Bulletin of the World Health Organization* 41, nos. 3-4-5 (1969): 492.

62 On disease ecology as a minor tradition within laboratory-dominated medicine during the twentieth century, see Anderson, "Natural Histories of Infectious Disease."

63 Staffan Müller-Wille and Hans-Jörg Rheinberger, *A Cultural History of Heredity* (Chicago: University of Chicago Press, 2012).

64 An early report of this research is in Frank Macfarlane Burnet, *Natural History of Infectious Disease*, 2nd ed. (Cambridge: Cambridge University Press, 1953). Burnet refers to "hybridization (or recombination)" of related flu viruses that "has recently been described from my own laboratory" (287–88).

65 Frank Fenner, "Frank Macfarlane Burnet 1899–1985," *Historical Records of Australian Science* 7, no. 1 (1987). During the 1940s, researchers at Caltech showed that viruses underwent mutations spontaneously and that different lines could be selected by changing conditions of the culture medium. In 1946 Edward Tatum and Joshua Lederberg showed that two lines of bacteria could exchange genetic material among themselves, a process of "recombination." Fenner himself was also involved in research on virus "ecology." On Fenner, see Warwick Anderson, "Nowhere to Run, Rabbit: The Cold-War Calculus of Disease Ecology," *History and Philosophy of the Life Sciences* 39, no. 2 (June 2017): 13, doi.org/10.1007/s40656-017-0140-7.

66 Edwin Kilbourne, "Recombination of Influenza A Viruses of Human and Animal Origin," *Science* 160, no. 3823 (1968): 74–76.

67 Robert Webster, "Antigenic Variation in Influenza Viruses, with Special Reference to Hong Kong Influenza," *Bulletin of the World Health Organization* 41, nos. 3–4–5 (1969): 483.

68 R. G. Webster, C. H. Campbell, and A. Granoff, "The '*In Vivo*' Production of 'New' Influenza Viruses: III. Isolation of Recombinant Influenza Viruses under Simulated Conditions of Natural Transmission," *Virology* 51, no. 1 (1973): 149–62. For swine "mixing vessel," see R. G. Webster et al., "Evolution and Ecology of Influenza A Viruses," *Microbiological Reviews* 56, no. 1 (1992): 179.

69 WHO Archives, ARCO10-3, Centralized Files, 3rd Generation, Sub-fonds 3, Box Z2-181-49, Webster to Cockburn, August 20, 1970.

70 R. G. Webster and C. H. Campbell, "Studies on the Origin of Pandemic Influenza: IV. Selection and Transmission of 'New' Influenza Viruses in Vivo," *Virology* 62, no. 2 (1974): 404–13.

71 On Lvov and the Moscow Laboratory of Ecology, see WHO Archives Box Z2-181-118(A); on Laver's seabird studies (with Australian National University), see WHO Archives Box Z2-181-144.

72 Webster and Campbell, "Studies on the Origin," 412.

73 WHO Archives Box Z2/180/11, "Ecology of Influenza: Framework for a Global Programme," 1982.

74 See the discussions among UN agencies included in WHO Archives, Box L2-308-302.

75 Graeme Laver, "Influenza Virus Surface Glycoproteins, Haemagglutinin, and Neuraminidase: A Personal Account," *Perspectives in Medical Virology* 7 (2002): 31–47.

76 WHO Archives, 12/181/12, CTS Agreement with the John Curtin School of Medical Research, Canberra, Australia, in respect of ecological study of influenza, January 12, 1973.

77 Laver, "Influenza Virus Surface Glycoproteins, Haemagglutinin, and Neuraminidase," 31–47.

78 WHO Archives, Box 12-133-1, Webster to John Skehal, June 10, 1981.

79 WHO Archives, Box 12-133-1.

80 Yuanji Guo et al., "Influenza Ecology in China," in *The Origin of Pandemic Influenza
Viruses: Proceedings of the International Workshop on the Molecular Biology and Ecology
of Influenza Virus Held in Peking, China, November 10–12, 1982*, ed. W. G. Laver (New
York: Elsevier, 1983).

81 Kennedy Shortridge and C. H. Stuart-Harris, "An Influenza Epicentre?," *Lancet* 2,
no. 8302 (1982): 812.

82 On Shortridge's laboratory as a "sentinel" for influenza pandemics, see Frédéric
Keck, "From Purgatory to Sentinel: 'Forms/Events' in the Field of Zoonoses,"
Cambridge Anthropology 32, no. 1 (2014): 47–61.

83 K. F. Shortridge, D. J. Alexander, and M. S. Collins, "Isolation and Properties of
Viruses from Poultry in Hong Kong Which Represent a New (Sixth) Distinct
Group of Avian Paroxymoviruses," *Journal of General Virology* 49 (1980): 255–62.

84 Shortridge, Alexander, and Collins, "Isolation and Properties," 260.

85 WHO Archives, Box 12-133-1, "Training Courses and Workshops on Influenza—
General," handwritten letter, Alan Kendal to Fakhry Assaad, May 11, 1981. Kendal
goes on to propose an alternative workshop format focused on training Chinese
scientists. "I agree with helping to train PRC scientists, and give them a sense of
perspective about molecular biology and influenza. This could be best done, I feel,
by a small group of 'western' scientists presenting a review/overview, of current
results, problems, and future objectives."

86 Shortridge and Stuart-Harris, "An Influenza Epicentre?," 812–13.

87 Shortridge and Stuart-Harris, "An Influenza Epicentre?," 812.

CHAPTER TWO: PATHOGENIC RESERVOIRS

1 Bruno Latour, *The Pasteurization of France* (Cambridge, MA: Harvard University
Press, 1988); Bruno Latour, *Science in Action: How to Follow Scientists and Engineers
through Society* (Cambridge, MA: Harvard University Press, 1987).

2 K. F. Shortridge et al., "Isolation and Characterization of Influenza A Viruses from
Avian Species in Hong Kong," *Bulletin of the World Health Organization* 55 (1977):
15–20.

3 Theresa MacPhail, *The Viral Network: A Pathography of the H1N1 Influenza Pandemic*
(Ithaca, NY: Cornell University Press, 2014), 35.

4 R. G. Webster et al., "Evolution and Ecology of Influenza A Viruses," *Microbiological
Reviews* 56, no. 1 (1992): 169.

5 On One Health, see Susan Craddock and Steve Hinchliffe, "One World, One
Health? Social Science Engagements with the One Health Agenda," *Social Science
and Medicine* 129 (March 2015): 1–4, doi.org/10.1016/j.socscimed.2014.11.016.

6 Warwick Anderson, "Natural Histories of Infectious Disease: Ecological Vision in
Twentieth-Century Biomedical Science," *Osiris* 19 (2004); 39–61.

7 Carlo Caduff, *The Pandemic Perhaps: Dramatic Events in a Public Culture of Danger* (Berkeley: University of California Press, 2015); A. M. Kleinman et al., "Asian Flus in Ethnographic and Political Context: A Biosocial Approach," *Anthropology and Medicine* 15, no. 1 (2008): 1–5, dx.doi.org/10.1080/13648470801918968.

8 Kennedy F. Shortridge, "Avian Influenza A Viruses of Southern China and Hong Kong: Ecological Aspects and Implications for Man," *Bulletin of the World Health Organization* 60, no. 1 (1982): 129–35. For additional background, see also Shortridge et al., "Isolation and Characterization," 15–20.

9 Kennedy F. Shortridge, "Pandemic Influenza: Application of Epidemiology and Ecology in the Region of Southern China to Prospective Studies," in *The Origin of Pandemic Influenza Viruses: Proceedings of the International Workshop on the Molecular Biology and Ecology of Influenza Virus Held in Peking, China, November 10–12, 1982*, ed. W. G. Laver (New York: Elsevier, 1983), 194.

10 Frédéric Keck, unpublished work.

11 Shortridge, "Pandemic Influenza," 195.

12 Shortridge, "Pandemic Influenza," 199.

13 Shortridge and Stuart-Harris, "An Influenza Epicentre?," 812.

14 Warwick Anderson writes of Macfarlane Burnet that he "used his interpretation of prewar ecology as a metaphoric resource rather than an analytic tool." Anderson, "Natural Histories," 51.

15 Paul Rabinow, *French DNA: Trouble in Purgatory* (Chicago: University of Chicago Press, 1999), 5.

16 "Development-by-accumulation" is Thomas Kuhn's description of "Whig" histories of science. "Paradigm shift" is his own alternative account of scientific development. Thomas S. Kuhn, *The Structure of Scientific Revolutions* (Chicago: University of Chicago Press, 1996).

17 Georges Canguilhem, "The History of Science," in *A Vital Rationalist: Selected Writings from Georges Canguilhem*, ed. Francois Delaporte (New York: Zone, 1994), 36.

18 Canguilhem, "The History of Science," 36.

19 Edward Gargan, "As Avian Flu Spreads, China Is Seen as Its Epicenter," *New York Times*, December 21, 1997.

20 L. D. Sims et al., "Avian Influenza in Hong Kong 1997–2002," *Avian Diseases* 47, no. 3 (September 2003): 832–38.

21 J. S. Peiris et al. "Re-emergence of Fatal Human Influenza A Subtype H5N1 Disease," *Lancet* 363, no. 9409 (2004): 617–19.

22 See Robert G. Wallace et al., "A Statistical Phylogeography of Influenza A H5N1," *Proceedings of the National Academy of Sciences of the United States of America* 104, no. 11 (2007): 4473–78.

23 Robert G. Wallace and Rodrick Wallace, *Neoliberal Ebola: Modeling Disease Emergence from Finance to Forest and Farm* (New York: Springer, 2016).

24 Shortridge, "Avian Influenza A Viruses," 133.

25 Pierre Legendre and Marie Josée Fortin, "Spatial Pattern and Ecological Analysis," *Vegetatio* 80, no. 2 (June 1989): 107–38.

26 "Marius Gilbert," *Prince Mahidol Award Conference 2013: A World United against Infectious Diseases: Cross-sectoral Solutions*, January 28, 2013, Bangkok, www .pmaconference.mahidol.ac.th/index.php?option=com_content&view=article&id =602&Itemid=195.

27 J. Slingenbergh et al., "Ecological Sources of Zoonotic Diseases," *Review of Science and Technology of the Office Internationale de Epizootiques* 23, no. 2 (2004): 469.

28 On the X-Ray, see Prasert Auewarakul, Wanna Hanchaoworakul, and Kumnuan Ungchusak, "Institutional Responses to Avian Influenza in Thailand: Control of Outbreaks in Poultry and Preparedness in the Case of Human-to-Human Transmission," *Anthropology and Medicine* 15, no. 1 (April 1, 2008): 61–67, doi:10.1080/13648470801919065.

29 Rachel M. Safman, "Avian Influenza Control in Thailand: Balancing the Interests of Different Poultry Producers," in *Avian Influenza: Science, Policy and Politics*, ed. Ian Scoones (New York: Routledge, 2010), 177.

30 Marius Gilbert et al., "Free-Grazing Ducks and Highly Pathogenic Avian Influenza, Thailand," *Emerging Infectious Diseases* 12, no. 2 (2006): 228.

31 Gilbert et al., "Free-Grazing Ducks," 230.

32 Marius Gilbert, "Spatial Epidemiology of Highly Pathogenic Avian Influenza (HPAI) in Thailand: Exploring the Use of Remote Sensing for Risk Assessment," in *Final Report of a Letter of Agreement between the Food and Agriculture Organization of the United Nations and the Université Libre de Bruxelles*, November 2005, 5.

33 Gilbert et al., "Free-Grazing Ducks," 232.

34 "Most rice fields in the eastern part of Thailand produce one crop per year whereas crops located in the central plains permit the production of two or even three crops per year. Single rice crop areas do associate with duck farming but count less ducks and a production cycle so as to match the period of rice harvest. In contrast, in double rice crop areas there is year round availability of post-harvest rice paddy fields, sustaining the low-input low-output free grazing duck farming system, and representing a very large proportion of the total duck numbers." Gilbert, "Spatial Epidemiology," 5.

35 Gilbert, "Spatial Epidemiology," 6.

36 Robert Peckham and Ria Sinha, "Satellites and the New War on Infection: Tracking Ebola in West Africa," *Geoforum* 80, Suppl. c (March 1, 2017): 24–38.

37 Chunglin Kwa, "Local Ecologies and Global Science: Discourses and Strategies of the International Geosphere-Biosphere Programme," *Social Studies of Science* 35, no. 6 (2005): 923–50.

38 In 2014 I learned about how these models are validated when I joined Xiao, a graduate student from Jiangxi Normal University, to conduct surveys of poultry farmers in the Poyang Lake region. He explained that he had recently surveyed rice farmers in the region. Equipped with a GPS device, he had asked a simple set of questions, including whether they planted one or two crops of rice per year. This geographically positioned survey data could then be compared with the maps produced using the index algorithms developed by Xiao Xiangming and others.

39 Marius Gilbert et al., "Avian Influenza, Domestic Ducks and Rice Agriculture in Thailand," *Agriculture, Ecosystems and Environment* 119, nos. 3–4 (2007): 412.

40 Marius Gilbert et al., "Mapping H5N1 Highly Pathogenic Avian Influenza Risk in Southeast Asia," *PNAS* 105, no. 12 (2008): 4770.

41 Marius Gilbert, "Spatial Modeling of H5N1: A Review," 3rd International Workshop on Community-Based Data Synthesis, Analysis and Modeling of Highly Pathogenic Avian Influenza H5N1 in Asia, November 2011, Beijing. As Gilbert and colleagues noted elsewhere, "In Indonesia, poultry census statistics are at a coarse spatial and temporal scale and the control of HPAI is correspondingly difficult. Remote sensing may help the construction of HPAI risk maps in Indonesia and Vietnam." Gilbert et al., "Avian Influenza, Domestic Ducks, and Rice Agriculture in Thailand," 414.

42 Scott H. Newman, Boripat Siriaroonat, and Xiangming Xiao, "A One Health Approach to Understanding Dynamics of Avian Influenza in Poyang Lake, China" (Kunming, China: Ecohealth, 2012).

43 Frédéric Keck, "From Purgatory to Sentinel: 'Forms/Events' in the Field of Zoonoses." *Cambridge Anthropology* 32, no. 1 (2014): 47–61.

44 WHO Archives, ARCO10-3, Centralized Files, 3rd Generation, Sub-fonds 3, Box Z2-180-11, Guo Yuanji to K. Esteves, February 7, 1985.

45 Gavin Smith, personal communication. This changed somewhat after 1997. In particular, Hong Kong University laboratory researcher Guan Yi initiated a parallel displacement to the one I discuss here. Born in mainland China, he trained with Robert Webster at St. Jude's before joining the Hong Kong lab. After 1997, he worked closely with Webster to develop new animal surveillance programs in Hong Kong. He then established a new lab in Shantou, a southern Guangdong city, at the University Medical College, and developed several virus sampling programs, including one at Poyang Lake.

CHAPTER THREE: LIVESTOCK REVOLUTIONS

1 FAOSTAT, www.fao.org/faostat/en/#data/QL.

2 Thomas P. Van Boeckel et al., "Modelling the Distribution of Domestic Ducks in Monsoon Asia," *Agriculture, Ecosystems and Environment* 141, nos. 3–4 (May 1, 2011): 373–80, doi.org/10.1016/j.agee.2011.04.013.

3 FAOSTAT. The U.S. produced 27 million ducks in 2017.

4 Peng Li et al., "Changes in Rice Cropping Systems in the Poyang Lake Region, China during 2004–2010," *Journal of Geographical Sciences* 22, no. 4 (2012): 653–68.

5 Hsing-Tsung Huang and Joseph Needham, *Science and Civilisation in China*, vol. 6, *Biology and Biological Technology* (Cambridge: Cambridge University Press, 2000). See also Bo Zhang et al., "从'稻田养鸭'到'稻鸭共生': 民国以来'稻田养鸭'技术的过渡与转型 - 以广东地区为中心." 农业考古 3 (2015).

6 Kennedy Shortridge and C. H. Stuart-Harris, "An Influenza Epicentre?," *Lancet* 2, no. 8302 (1982): 812–13.

7 Marius Gilbert et al., "Free-Grazing Ducks and Highly Pathogenic Avian Influenza, Thailand," *Emerging Infectious Diseases* 12, no. 2 (2006): 227–34.

8 On zoonotic diagrams, see Christos Lynteris, "Zoonotic Diagrams: Mastering and Unsettling Human–Animal Relations," *Journal of the Royal Anthropological Institute* 23, no. 3 (2017): 463–85. Here I am using the term more expansively to discuss not only actual diagrammatic artifacts but more broadly hypothetical models of the interspecies relations that drive the emergence of influenza pandemics (models that are often but not always formalized as diagrams).

9 Yong Wang et al., "Risk Factors for Infectious Diseases in Backyard Poultry Farms in the Poyang Lake Area, China," *PLoS ONE* 8, no. 6 (2013).

10 Lynteris, "Zoonotic Diagrams."

11 FAO, "Avian Influenza: Stop the Risk for Humans and Animals at Source," donor appeal, circa 2005, accessed April 1, 2019, www.fao.org/avianflu/documents /donor.pdf.

12 FAO and OIE, "A Global Strategy for the Progressive Control of Highly Pathogenic Avian Influenza (HPAI)," November 2005, 22, www.fao.org/avianflu/documents /hpaiglobalstrategy31oct05.pdf. Underlined in original. The report further specifies that "source of infection" "refers predominantly to the smallholder poultry sector and the domestic duck population, a major carrier host reservoir."

13 Phil Harris, "Avian Influenza: An Animal Health Issue," August 2006, www.fao.org /avianflu/en/issue.html, emphasis in original.

14 Samimian-Darash contrasts *potential* and *possible* uncertainty as uninsurable to insurable risk. Limor Samimian-Darash, "Governing Future Potential Biothreats: Toward an Anthropology of Uncertainty," *Current Anthropology* 54, no. 1 (2013): 1–22. On uninsurable risk, see also Stephen J. Collier and Andrew Lakoff, "Vital Systems Security: Reflexive Biopolitics and the Government of Emergency," *Theory, Culture and Society* 32, no. 2 (2015): 19–51.

15 For example, Jessica A. Belser et al., "Use of Animal Models to Understand the Pandemic Potential of Highly Pathogenic Avian Influenza Viruses," *Advances in Virus Research* 73 (2009): 55–97, doi.org/10.1016/S0065-3527(09)73002-7.

16 For some examples, see Andrew Lakoff, "The Generic Biothreat, or, How We Became Unprepared," *Cultural Anthropology* 23, no. 3 (2008): 399–428; Emily Waller, Mark Davis, and Niamh Stephenson, "Australia's Pandemic Influenza 'Protect' Phase: Emerging out of the Fog of Pandemic," *Critical Public Health* 26, no. 1 (January 1, 2016): 99–113, doi.org/10.1080/09581596.2014.926310; Lyle Fearnley, "Signals Come and Go: Syndromic Surveillance and Styles of Biosecurity," *Environment and Planning A* 40, no. 7 (2008): 1615–32; Carlo Caduff, *The Pandemic Perhaps: Dramatic Events in a Public Culture of Danger* (Berkeley: University of California Press, 2015); Collier and Lakoff, "Vital Systems Security"; Samimian-Darash, "Governing Future Potential Biothreats."

17 Lakoff argues that the ethical stance of global health security is "self-protection" against "emerging infectious diseases that threaten wealthy countries." Andrew Lakoff, "Two Regimes of Global Health," *Humanity* 1, no. 1 (2010): 59–79.

18 For example, Brian Wynne, "May the Sheep Safely Graze? A Reflexive View of the Expert–Lay Divide," in *Risk, Environment and Modernity: Towards a New Ecology*, ed. Scott Lash, Bronislaw Szerszynski, and Brian Wynne (London: SAGE, 1996).

19 FAO, "Avian Influenza Control and Eradication: FAO's Proposal for a Global Programme," Rome, March 2006, 13.

20 FAO, "Recommendations on the Prevention, Control and Eradication of Highly Pathogenic Avian Influenza (HPAI) in Asia (Proposed with the Support of OIE)," Rome, September 2004, 6.

21 D. E. Swayne and D. L. Suarez, "Highly Pathogenic Avian Influenza," *Review of Science and Technology of the Office Internationale de Epizootiques* 19, no. 2 (2000): 473.

22 Abigail Woods, *A Manufactured Plague: The History of Foot-and-Mouth Disease in Britain* (London: EARTHSCAN, 2004).

23 Eunha Shim and Alison P. Galvani, "Evolutionary Repercussions of Avian Culling on Host Resistance and Influenza Virulence," *PLoS ONE* 4, no. 5 (May 11, 2009): e5503.

24 See, for example, Mei Fuchun, *Research on the Government's Use of Culling Compensation Policy in Response to the Avian Influenza Crisis* 《政府应对禽流感突发事件的扑杀补偿政策研究》 中国农业出版社, 2011.

25 I discuss the immunization program in detail in chapter 5.

26 The FAO provided guidance to national governments on the use of vaccination to control and prevent outbreaks. However, the FAO warned that vaccination "should not be considered as a permanent measure" and called on national governments to develop an immediate "exit strategy." FAO, "Recommendations," 8.

27 FAO documents and discussions first deployed the concept of biosecurity in the early 2000s. At the time the FAO was not especially concerned with avian and pandemic influenza, and the promotion of biosecurity reflected an awareness of growing global trade in agricultural products and, more specifically, the emergence of food-safety regulation as a key component of free trade agreements. In 2002 a group of "international experts" met in Rome for an expert consultation on biosecurity in food and agriculture. Biosecurity was new to the FAO: as a subsequent report on the meeting acknowledges, "'*Biosecurity*' is a relatively new concept, and a term with still evolving meanings. The concepts behind *Biosecurity* are relatively new, and there is therefore some confusion in terminology." Indeed, several participants in the 2002 Rome meeting had concerns about the possible untranslatability of the concept of biosecurity from English into other languages. In the end, *Biosecurity* (capitalized and italicized, but retained in English in all translated texts) was selected as the term that "best describes the concept as used by FAO." In this period the FAO focused on risk management at a largely regulatory and transnational level: "Broadly speaking *Biosecurity* describes the concept, process and objective of managing—in a holistic manner—biological risks associated with food and agriculture." The agenda of the 2003 technical consultation in Bangkok included discussions of risk analysis as a "unifying concept"; information access and exchange, particularly at a transnational scale; and capacity building for "biological risk management." When the FAO became involved in avian flu after the reemergence of the highly pathogenic H5N1 virus in 2004, however, biosecurity became a key conceptual framework for planning interventions into the epicenter. In the first major report the FAO issued on the avian influenza

crisis in 2004, the Animal Health Service called for "enhanced biosecurity of poultry farms and associated premises" as a central recommendation. See FAO, "Recommendations."

28 The wide use of the term *biosecurity* sometimes obscures the fact that it remains highly polysemous, despite several efforts to "unify" terminology (including by the FAO). The term is deployed to refer to agricultural farm management, regulation of trade in agricultural products, invasive species, and preparedness for biological terrorism and emerging diseases. See Stephen Hinchliffe and Nick Bingham, "Securing Life: The Emerging Practices of Biosecurity," *Environment and Planning* A 40, no. 7 (2008): 1534–51; Andrew Lakoff and Stephen J. Collier, *Biosecurity Interventions: Global Health and Security in Question* (New York: Columbia University Press, 2008); Nancy N. Chen and Lesley Alexandra Sharp, *Bioinsecurity and Vulnerability* (Santa Fe, NM: School for Advanced Research Press, 2014).

29 FAO, "Biosecurity for Highly Pathogenic Avian Influenza: Issues and Options," Rome, 2008, 1.

30 FAO, "Biosecurity for Highly Pathogenic Avian Influenza," 1.

31 See Natalie Porter, "Bird Flu Biopower: Strategies for Multispecies Coexistence in Việt Nam," *American Ethnologist* 40, no. 1 (2013): 132–48.

32 FAO, "Biosecurity for Highly Pathogenic Avian Influenza," 22.

33 FAO, "Recommendations," 7.

34 FAO, "Avian Influenza Control and Eradication: FAO's Proposal for a Global Programme," Rome, March 2006, 13.

35 FAO, "Recommendations." In a presentation at the February 2004 FAO/OIE joint conference on avian influenza in Bangkok, Joseph Annelli of the USDA presented a three-part typology that could perhaps be a progenitor: (i) industrial integrated systems with established biosecurity, (ii) commercial nonintegrated systems of poultry production with minimal biosecurity, and (iii) village production of native chickens and fighting cocks. FAO/OIE Emergency Regional Meeting on Avian Influenza Control in Animals in Asia, February 26–28, 2004, Bangkok.

36 Paul Forster and Olivier Charnoz, "Producing Knowledge in Times of Health Crises: Insights from the International Response to Avian Influenza in Indonesia," *Revue d'anthropologie des connaissances* 7, no. 1 (2013): 112–44. Forster and Charnoz note that the origin of this typology is obscure but is perhaps rooted in the FAO livestock-development schemes from the 1960s and 1970s. They also point out that there have been slight variations in the formulation of this typology over time.

37 Olaf Thieme and Emanuelle Guerne Bleich, "Poultry Sector Restructuring for Disease Control: Initial Thoughts," FAO Animal Production Service, Rome, April 2007, www.fao.org/docs/eims/upload/239034/ai291e.pdf. Notably, when I visited Vietnam on an FAO project in 2014, live-poultry markets remained the primary source of poultry meat, and FAO officials were still struggling to figure out how to increase the percentage of slaughtered poultry sold to consumers.

38 John Law, "Disaster in Agriculture: Or Foot and Mouth Mobilities," *Environment and Planning* A 38, no. 2 (February 1, 2006): 227–39, doi.org/10.1068/a37273.

39 S. Hinchliffe et al., "Biosecurity and the Topologies of Infected Life: From Borderlines to Borderlands," *Transactions of the Institute of British Geographers* 38, no. 4 (2013): 534.

40 Elizabeth C. Dunn, "Trojan Pig: Paradoxes of Food Safety Regulation," *Environment and Planning A* 35 (2003): 1493–511.

41 Mike Davis, *The Monster at Our Door: The Global Threat of Avian Flu* (New York: New Press, 2005); Celia Lowe, "Viral Clouds: Becoming H5N1 in Indonesia," *Cultural Anthropology* 25, no. 4 (2010): 625–49; Robert Wallace, *Big Farms Make Big Flu: Dispatches on Infectious Disease, Agribusiness, and the Nature of Science* (New York: Monthly Review Press, 2016).

42 In the English language a similar linguistic move is made in (informal) business or economics discourse when one refers to a business that "has not achieved scale," which actually means not achieving *large enough* scale (sufficient size of customer base, for instance) for a particular business model to work effectively.

43 See, for instance, P. Cherry and T. R. Morris, *Domestic Duck Production: Science and Practice* (Cambridge, MA: CABI, 2008).

44 Jikun Huang, Xiaobing Wang, and Guangguang Qiu, *Small-Scale Farmers in China in the Face of Modernisation and Globalisation* (London: IIED/HIVOS, 2012), 16.

45 Cited in Qian Forrest Zhang and John A. Donaldson, "The Rise of Agrarian Capitalism with Chinese Characteristics: Agricultural Modernization, Agribusiness, and Collective Land Rights," *China Journal* 60 (2008): 28.

46 Phillip C. C. Huang, "China's New-Age Small Farms and Their Vertical Integration: Agribusiness or Co-ops?," *Modern China* 32, no. 2 (2011): 107–34.

47 Huang, "China's New-Age Small Farms," 111.

48 Although farmer co-operatives (*hezuoshe*) are another widely cited mechanism of vertical integration in contemporary China, recent research suggests that most, if not all, farmer co-ops are either frauds or failures. That is, "genuine" co-ops that may have existed at one time no longer do so, probably because they failed or they are no longer genuine. Most existing co-ops are merely "shells" for receiving lucrative government subsidies or used by for-profit agribusiness to access rural farmland (for instance, in the *gongsi* + *hezuoshe* + *nonghu* model). Zhanping Hu, Qian Forrest Zhang, and John A. Donaldson, "Farmers' Cooperatives in China: A Typology of Fraud and Failure," *China Journal* 78 (2017): 1–24.

49 Susanne Lingohr, "Rural Households, Dragon Heads and Associations: A Case Study of Sweet Potato Processing in Sichuan Province," *China Quarterly* 192 (2007): 898–914.

50 Mindi Schneider, "Dragon Head Enterprises and the State of Agribusiness in China," *Journal of Agrarian Change* 17, no. 1 (2017): 3–21.

51 "前后当了38年的兽医," 南方网2009年08月31日.

52 Yong Wang et al., "Risk Factors for Infectious Diseases in Backyard Poultry Farms in the Poyang Lake Area, China," *PLoS ONE* 8, no. 6 (June 20, 2013): 2, doi.org/10 .1371/journal.pone.0067366.

53 However, the difference from the pre-Communist sideline farming that I described above should not be forgotten. In the Mao era the "sideline" was what the household could eke out on the side of their work for the collective.

54 Jonathan Unger, *The Transformation of Rural China* (London: Routledge, 2016), 21.

55 第三十七条国家支持农村集体经济组织、农民和畜牧业合作经济组织建立畜禽养殖场、养殖小区，发展规模化、标准化养殖。乡（镇）土地利用总体规划应当根据本地实际情况安排畜禽养殖用地。农村集体经济组织、农民、畜牧业合作经济组织按照乡（镇）土地利用总体规划建立的畜禽养殖场、养殖小区用地按农业用地管理。畜禽养殖场、养殖小区用地使用权期限届满，需要恢复为原用途的，由畜禽养殖场、养殖小区土地使用权人负责恢复。在畜禽养殖场、养殖小区用地范围内需要兴建永久性建（构）筑物，涉及农用地转用的，依照《中华人民共和国土地管理法》的规定办理。中国人民共和国畜牧法。2006. 第四章，第三十七条.。

56 Susanne Kerner, Cynthia Chou, and Morten Warmind, *Commensality: From Everyday Food to Feast* (New York: Bloomsbury, 2015).

57 Marilyn Strathern, "Eating (and Feeding)," *Cambridge Anthropology* 30, no. 2 (2012): 1–14.

58 Edmund Leach, "Anthropological Aspects of Language: Animal Categories and Verbal Abuse," *Anthrozoös* 2, no. 3 (1989): 161. See also Stanley J. Tambiah, "Animals Are Good to Think and Good to Prohibit," *Ethnology* 8, no. 4 (1969): 423–59.

59 C. Fausto and L. Costa, "Feeding (and Eating): Reflections on Strathern's 'Eating (and Feeding),'" *Cambridge Anthropology* 31, no. 1 (2013): 157.

60 David Goodman, Bernardo Sorj, and John Wilkinson, *From Farming to Biotechnology: A Theory of Agro-industrial Development* (New York: Basil Blackwell, 1987).

61 Ke Bingsheng and Yijun Han, "Poultry Sector in China: Structural Change during the Past Decade and Future Trends," Food and Agriculture Organization of the United Nations, 2008.

62 Leach, "Anthropological Aspects of Language," uses this language of "bringing close" and "pushing farther."

63 William Boyd and Michael Watts, "Agro-industrial Just-in-Time: The Chicken Industry and Postwar American Capitalism," in *Globalising Food: Agrarian Questions and Global Restructuring*, ed. David Goodman and Michael Watts (Hoboken, NJ: Taylor and Francis, 2013).

64 Cherry and Morris, *Domestic Duck Production*.

65 Ke Bingsheng and Yijun Han, "Poultry Sector in China."

66 Heather Paxson, *The Life of Cheese: Crafting Food and Value in America* (Berkeley: University of California Press, 2012); Brad Weiss, *Real Pigs: Shifting Values in the Field of Local Pork* (Durham, NC: Duke University Press, 2016). On local food in China, see Jakob Klein, "Connecting with the Countryside? 'Alternative' Food Movements with Chinese Characteristics," in *Ethical Eating in the Postsocialist and Socialist World*, ed. Yuson Jung, Jakob A. Klein, and Melissa L. Caldwell (Berkeley: University of California Press, 2014).

67 James C. Scott, *The Moral Economy of the Peasant: Rebellion and Subsistence in Southeast Asia* (New Haven, CT: Yale University Press, 1976), 5.

68 Indeed, until at least the 1980s chicken or duck was primarily consumed only during festivals or on other special occasions. See Tik-sang Liu, "Custom, Taste and Science: Raising Chickens in the Pearl River Delta Re-

gion, South China," *Anthropology and Medicine* 15, no. 1 (April 1, 2008): 7–18, doi.10.1080/13648470801918992.

69 The word *tu* literally means "soil." The word is sometimes used derisively to refer to "backward" peasants, but it is also used in a highly positive sense to refer to small-scale, rural poultry farming (*tu ji*, "soil chicken"). Jacob Klein, in a book chapter on the emergence of local and organic food movements in Yunnan, China, translates *tu* as "earthy." See Klein, "Connecting with the Countryside?"

70 Anna Lora-Wainwright, "Of Farming Chemicals and Cancer Deaths: The Politics of Health in Contemporary Rural China," *Social Anthropology* 17, no. 1 (2009): 56–73; Anna Lora-Wainwright, *Fighting for Breath: Living Morally and Dying of Cancer in a Chinese Village* (Honolulu: University of Hawai'i Press, 2013).

71 Goncalo Santos, "Rethinking the Green Revolution in South China: Technological Materialities and Human–Environment Relations," *East Asian Science, Technology and Society* 5, no. 4 (2011): 495. Elsewhere in China, rural Henan villagers claim that farm chemicals increase production yet lament that their use causes the soil to "lose strength." Still, they use these chemicals as much as they can, a practice that Lai Lili suggests has "much to do with their deepened dependence on a cash economy." Lili, "Everyday Hygiene in Rural Henan," *Positions* 22, no. 3 (2014): 635–59.

72 Ellen Oxfeld, *Drink Water, but Remember the Source: Moral Discourse in a Chinese Village* (Berkeley: University of California Press, 2010).

73 On investment and risk in poultry farming during a pandemic flu crisis, see also Siwi Padmawati and Mark Nichter, "Community Response to Avian Flu in Central Java, Indonesia," *Anthropology and Medicine* 15, no. 1 (April 1, 2008): 31–51, doi.10.1080/13648470801919032.

74 "For instance grape after being pressed must ferment awhile and then rest for some time in order to reach a certain degree of perfection. In many branches of industry the product must pass through a drying process, for instance in pottery, or be exposed to certain conditions in order to change its chemical properties, as for instance in bleaching. Winter grain needs about nine months to mature. Between the time of sowing and harvesting the labor-process is almost entirely suspended. In timber-raising, after the sowing and the incidental preliminary work are completed, the seed requires about 100 years to be transformed into a finished product and during all that time it stands in comparatively very little need of the action of labor." Karl Marx, *Capital* (London: Penguin, 1978), 2:316.

75 See also Tim Ingold, "Making Things, Growing Plants, Raising Animals, and Bringing Up Children," in Ingold, *Perception of the Environment: Essays on Livelihood, Dwelling, and Skill* (London: Routledge, 2000).

76 Compensation rates are notoriously too low, often prompting farmers to attempt illegal sale of sick or condemned birds. See Mei Fuchun, *Research on the Government's Use of Culling Compensation Policy in Response to the Avian Influenza Crisis*《政府应对禽流感突发事件的扑 杀补偿政策研究》中国农业出版社, 2011.

77 Pierre Bourdieu, *The Logic of Practice* (Stanford, CA: Stanford University Press, 1990).

78 Edmund Leach, *Political Systems of Highland Burma: A Study of Kachin Social Structure* (Boston: Beacon, 1965).

79 Bronislaw Malinowski, *Coral Gardens and Their Magic*, vol. 1, *Soil-Tilling and Agricultural Rites in the Trobriand Islands* (Bloomington: Indiana University Press, 1965), 56.

80 Alfred Gell, "The Technology of Enchantment and the Enchantment of Technology," in *The Art of Anthropology: Essays and Diagrams*, edited by Eric Hirsch (London: Athlone, 1999), 182.

81 Malinowski, *Coral Gardens and Their Magic*, 76.

82 See chapter 6 for a detailed discussion of the duck doctor's practice.

83 Bu Liping, "Anti-malaria Campaigns and the Socialist Reconstruction of China, 1950–1980," *East Asian History* 39 (2014): 117–30.

84 Miriam Gross, *Farewell to the God of Plague: Chairman Mao's Campaign to Deworm China* (Oakland: University of California Press, 2016).

85 Dropping the word for "needle" (*zhen*) from the more complete phrase *da yufangzhen*.

86 For example, a 2012 FAO report notes that "many small operators were put out of business." FAO, *Lessons from HPAI: A Technical Stockpiling of Outputs, Outcomes, Best Practices, and Lessons Learned from the Fights against Highly Pathogenic Avian Influenza in Asia 2005–2011* (FAO Animal Production and Health Paper no. 176, Rome, 2013, 50).

87 Bingsheng Ke and Yijun Han, "Poultry Sector in China."

88 Chaoping Xie and Mary A. Marchant, "Supplying China's Growing Appetite for Poultry," *International Food and Agribusiness Management Review* 18, Special Issue A (2015): 123.

89 *Liugan*, she explained, did not result in the death of the entire flock but simply caused the birds to eat less feed and produce fewer eggs. The drop in production was why this could cause the farmer to go broke.

90 On the biocommunicability of the pandemic threat, see Charles L. Briggs and Mark Nichter, "Biocommunicability and the Biopolitics of Pandemic Threats," *Medical Anthropology* 28, no. 3 (2009): 189–98.

91 For a relevant account in another part of China, see Letian Zhang and Tianshu Pan, "Surviving the Crisis: Adaptive Wisdom, Coping Mechanisms and Local Responses to Avian Influenza Threats in Haining, China," *Anthropology and Medicine* 15, no. 1 (2008): 19–30.

92 Julien Cappelle et al., "Risks of Avian Influenza Transmission in Areas of Intensive Free-Ranging Duck Production with Wild Waterfowl," *EcoHealth* 11 (2014): 109–19.

93 Note there are important differences between insurance and free grazing. Insurance manages market uncertainty while *not* reducing dependence on market-based sources of feed. As a result, insurance would in theory enable farmers to embark on larger scales of farming and take on higher risk. By contrast, free grazing reduces exposure to risk by reducing dependence on market feeds.

94 FAO, *Lessons from HPAI*, 55.

CHAPTER FOUR: WILD GOOSE CHASE

1 See, for example, Charis Thompson, "When Elephants Stand for Competing Philosophies of Nature: Amboseli National Park, Kenya," in *Complexities: Social*

Studies of Knowledge Practice, ed. John Law and Annemarie Mol (Durham, NC: Duke University Press, 2006).

2 The foundational works are Karin Knorr-Cetina, *The Manufacture of Knowledge: An Essay on the Constructivist and Contextual Nature of Science* (Oxford: Pergamon, 1981); Bruno Latour and Steve Woolgar, *Laboratory Life: The Construction of Scientific Facts* (Princeton, NJ: Princeton University Press, 1986); Sharon Traweek, *Beamtimes and Lifetimes: The World of High Energy Physicists* (Cambridge, MA: Harvard University Press, 1992). For review, see Karin Knorr-Cetina, "Laboratory Studies: The Cultural Approach to the Study of Science," in *Handbook of Science and Technology Studies*, ed. Sheila Jasanoff, Gerald E. Markle, James C. Peterson, and Trevor Pinch (Thousand Oaks, CA: SAGE, 1995).

3 See, for example, Steven Shapin and Simon Schaffer, *Leviathan and the Air-Pump: Hobbes, Boyle, and the Experimental Life* (Princeton, NJ: Princeton University Press, 1985); Andrew Cunningham, "Transforming Plague: The Laboratory and the Identity of Infectious Disease," in *The Laboratory Revolution in Medicine*, ed. Andrew Cunningham and Perry Williams (Cambridge: Cambridge University Press, 2002).

4 Karin Knorr-Cetina, "The Couch, the Cathedral, and the Laboratory: On the Relationship between Experiment and Laboratory in Science," in *Science as Practice and Culture*, ed. Andrew Pickering (Chicago: University of Chicago Press, 2010).

5 Knorr-Cetina, "The Couch, the Cathedral, and the Laboratory."

6 Bruno Latour, *Pandora's Hope: Essays on the Reality of Science Studies* (Cambridge, MA: Harvard University Press, 1999).

7 David N. Livingstone, *Putting Science in Its Place: Geographies of Scientific Knowledge* (Chicago: University of Chicago Press, 2013).

8 Jeremy Vetter, "Introduction," in *Knowing Global Environments: New Historical Perspectives on the Field Sciences*, ed. Jeremy Vetter (New Brunswick, NJ: Rutgers University Press, 2010).

9 Henrika Kuklick and Robert E. Kohler, "Science in the Field," *Osiris* 11 (1996): 1–265.

10 Robert Kohler, "Prospects," in *Knowing Global Environments: New Historical Perspectives on the Field Sciences*, ed. Jeremy Vetter (New Brunswick, NJ: Rutgers University Press, 2010).

11 Robert Kohler, *Landscapes and Labscapes: Exploring the Lab–Field Border in Biology* (Chicago: University of Chicago Press, 2002), 11.

12 "Unprecedented event" is from Hans-Jörg Rheinberger, *Toward a History of Epistemic Things: Synthesizing Proteins in the Test Tube* (Stanford, CA: Stanford University Press, 1997).

13 Jiao Li, "In China's Backcountry, Tracking Lethal Bird Flu," *Science* 330, no. 6002 (2010): 313.

14 J. Y. Takekawa et al., "Victims and Vectors: Highly Pathogenic Avian Influenza H5N1 and the Ecology of Wild Birds," *Avian Biology Research* 3, no. 2 (2010): 51–73.

15 H Chen et al., "Establishment of Multiple Sublineages of H5N1 Influenza Virus in Asia: Implications for Pandemic Control," *Proceedings of the National Academy of Sciences of the United States of America* 103, no. 8 (2006): 2849.

16 Li, "In China's Backcountry," 313.
17 Scott H. Newman, Boripat Siriaroonat, and Xiangming Xiao, "A One Health Approach to Understanding Dynamics of Avian Influenza in Poyang Lake, China" (Kunming, China: Ecohealth, 2012).
18 Takekawa et al., "Victims and Vectors," 3.
19 Takekawa et al., "Victims and Vectors," 15.
20 Takekawa et al., "Victims and Vectors," 4.
21 Julien Cappelle et al., "Risks of Avian Influenza Transmission in Areas of Intensive Free-Ranging Duck Production with Wild Waterfowl," *EcoHealth* 11 (2014): 109–19.
22 Cappelle et al., "Risks of Avian Influenza Transmission," 110.
23 Michael J. Hathaway, *Environmental Winds: Making the Global in Southwest China* (Berkeley: University of California Press, 2013).
24 《鄱阳湖研究》编委会. 鄱阳湖研究 = *Studies on Poyang Lake*. 上海: 上海科学技术出版社: 新華書店上海发行所发行, 1988.
25 Tan Chen, "Yangzhi Dayan, Qianjing Guangkuo," *Countryside-Agriculture Peasants* 10 (1999): 21. See also Sichuan Nanyang Special Type Economic Animal and Plant Association, "Xinxing Tezhong Yangzhi: Dayan," *Rural New Technology* 10 (1999).
26 Yuming Li, "Yesheng Dongwu yu Tezhong Yangzhi," *Peasant Daily*, March 15, 2001. See also Linden J. Ellis and Jennifer L. Turner, "Where the Wild Things Are . . . Sold," *China Environment Series* 9 (2007): 131–34, www.wilsoncenter.org/publication/full-publication-1.
27 Karl Taro Greenfeld, "Wild Flavor," *Paris Review* 47, no. 175 (2005): 7.
28 Mei Zhan, "Civet Cats, Fried Grasshoppers, and David Beckham's Pajamas: Unruly Bodies after SARS," *American Anthropologist* 107, no. 1 (2005): 31–42; Mei Zhan, "Wild Consumption: Privatizing Responsibilities in the Time of SARS," in *Privatizing China: Socialism from Afar*, ed. Aihwa Ong and Li Zhang (Ithaca, NY: Cornell University Press, 2008).
29 Ke Bingsheng and Yijun Han, "Poultry Sector in China: Structural Change during the Past Decade and Future Trends," Food and Agriculture Organization of the United Nations, 2008.
30 Tim Ingold, *Hunters, Pastoralists and Ranchers: Reindeer Economies and Their Transformations* (Cambridge: Cambridge University Press, 1980).
31 Yuming Li, "Yesheng Dongwu yu Tezhong Yangzhi."
32 See Yan Yunxiang, "Food Safety and Social Risk in Contemporary China," *Journal of Asian Studies* 71, no. 3 (2012): 705–29.
33 Chris Coggins, *The Tiger and the Pangolin: Nature, Culture and Conservation in China* (Honolulu: University of Hawai'i Press, 2004).
34 Ann Anagnost, *National Past-Times: Narrative, Representation, and Power in Modern China* (Durham, NC: Duke University Press, 1997).
35 On discourses of creativity in contemporary China, see Winnie Won Yin Wong, *Van Gogh on Demand: China and the Readymade* (Chicago: University of Chicago Press, 2014).
36 Luc Boltanski, Ève Chiapello, and Gregory Elliott, *The New Spirit of Capitalism* (London: Verso, 2018).

37 On "low quality," see Ann Anagnost, "The Corporeal Politics of Quality (*Suzhi*)," *Public Culture* 16, no. 2 (2004): 189–208. I discuss the concept at greater length in chapter 6.

38 Paul Rabinow, "Artificiality and Enlightenment," in Rabinow, *Essays on the Anthropology of Reason* (Princeton, NJ: Princeton University Press, 1996), 91–111.

39 See, for example, Andrew Pickering, *The Mangle of Practice: Time, Agency and Science* (Chicago: University of Chicago Press, 1995).

40 Hans-Jörg Rheinberger, *Toward a History of Epistemic Things: Synthesizing Proteins in the Test Tube* (Stanford, CA: Stanford University Press, 1997), 134.

41 Rheinberger, *Toward a History*, 28.

42 Rheinberger, *Toward a History*, 11.

43 Rheinberger, *Toward a History*, 134.

44 Darell Whitworth et al., "Wild Birds and Avian Influenza: An Introduction to Applied Research and Field Sampling Techniques," FAO Animal Production and Health Manual, Rome, 2007, 27–28.

45 Scott H. Newman, Boripat Siriaroonat, and Xiangming Xiao, "A One Health Approach to Understanding Dynamics of Avian Influenza in Poyang Lake, China," Kunming, China, Ecohealth, 2012.

46 Knorr-Cetina, "The Couch, the Cathedral, and the Laboratory," 116. See also Etienne Benson, *Wired Wilderness: Technologies of Tracking and the Making of Modern Wildlife* (Baltimore: Johns Hopkins University Press, 2010).

47 Rheinberger, *Toward a History*, 134.

48 Paul Rabinow and William M. Sullivan, *Interpretive Social Science: A Second Look* (Berkeley: University of California Press, 1987).

49 Changqing Ding, "Farmed wild ducks and geese: the missing link in understanding the diversity and epidemiology of pathogens transmitted among poultry, wildlife, and people." The 3rd International Workshop on Community-based Data Synthesis, Analysis and Modeling of Highly Pathogenic Avian Influenza H5N1 in Asia, November 14–15, 2011, Beijing, China.

CHAPTER FIVE: AFFINITY AND ACCESS

1 Andrew Lakoff, "Two Regimes of Global Health," *Humanity* 1, no. 1 (2010): 59–79; Theodore M. Brown, Marcos Cueto, and Elizabeth Fee, "The World Health Organization and the Transition from 'International' to 'Global' Public Health," *American Journal of Public Health* 96, no. 1 (January 2006): 62, doi.org/10.2105/AJPH .2004.050831.

2 "WHO Blasts China for Withholding Bird Flu Samples," *New Scientist* (November 3, 2006), www.newscientist.com/article/dn10439-who-blasts-china-for-withholding -bird-flu-samples.

3 Amy Hinterberger and Natalie Porter, "Genomic and Viral Sovereignty: Tethering the Materials of Global Biomedicine," *Public Culture* 27, no. 2–76 (2015): 361–86; David P. Fidler, "Influenza Virus Samples, International Law, and Global Health Diplomacy," *Emerging Infectious Diseases* 14, no. 1 (2008): 88–94; Theresa

MacPhail, "The Politics of Bird Flu: The Battle over Viral Samples and China's Role in Global Public Health," *Journal of Language and Politics* 8, no. 3 (2009): 456–75.

4 For an exception, see Niamh Stephenson, "Emerging Infectious Disease/Emerging Forms of Biological Sovereignty," *Science, Technology, and Human Values* 36, no. 5 (2010), doi.10.1177/0162243910388023.

5 Wen-Hua Kuo, "The Voice on the Bridge: Taiwan's Regulatory Engagement with Global Pharmaceuticals," *East Asian Science, Technology and Society* 3, no. 1 (March 1, 2009): 51–72, doi.org/10.1215/s12280-008-9066-1.

6 Vincanne Adams, Thomas E. Novotny, and Hannah Leslie state that global health diplomacy refers to an "emerging field that addresses the dual goals of improving global health and bettering international relations." Adams, Novotny, and Leslie, "Global Health Diplomacy," *Medical Anthropology* 27, no. 4 (2008): 316. On scientific diplomacy, see also Michael M. J. Fischer, "Biopolis: Asian Science in the Global Circuitry," *Science, Technology and Society* 18, no. 3 (2013): 381–406.

7 Didier Fassin and Mariella Pandolfi, "Introduction: Military and Humanitarian Government in the Age of Intervention," in *Contemporary States of Emergency: The Politics of Military and Humanitarian Interventions*, ed. Didier Fassin and Mariella Pandolfi (Cambridge, MA: Zone, 2010).

8 Peter Redfield, *Life in Crisis: The Ethical Journey of Doctors without Borders* (Berkeley: University of California Press, 2014). On "humanitarian biomedicine" and "global health security" as distinct regimes of global health, see Andrew Lakoff, *Unprepared: Global Health in a Time of Emergency* (Oakland: University of California Press, 2017).

9 "Ethical journey" is from Redfield, *Life in Crisis*.

10 See Pierre Bourdieu on land and matrimonial strategies in *The Logic of Practice* (Stanford, CA: Stanford University Press, 1990), 147–61.

11 Hualan Chen and Zhigao Bu, "Development and Application of Avian Influenza Vaccines in China," in *Vaccines for Pandemic Influenza*, ed. Richard W. Compans and Walter A. Orenstein (Berlin: Springer Berlin Heidelberg, 2009), 153–62, doi.org/10.1007/978-3-540-92165-3_7; Mara Hvistendahl, "Veterinarian-in-Chief," *Science* 341, no. 6142 (2013): 122–25; 田国彬. "H5 亚型禽流感灭活疫苗的研制及应用回顾与展望." 中国农业科学 40, no. 1 (2007): 444–48.

12 Quoted in Hvistendahl, "Veterinarian-in-Chief," 123.

13 Hvistendahl, "Veterinarian-in-Chief," 123.

14 Chen and Bu, "Development and Application of Avian Influenza Vaccines," 154.

15 Hvistendahl, "Veterinarian-in-Chief," 123.

16 Erich Hoffmann et al., "Eight-Plasmid System for Rapid Generation of Influenza Virus Vaccines," *Vaccine* 20, nos. 25–26 (August 19, 2002): 3165–70.

17 S. Pleschka et al., "A Plasmid-Based Reverse Genetics System for Influenza A Virus," *Journal of Virology* 70, no. 6 (June 1996): 4188–92.

18 Qiao Chuanling et al., "Vaccines Developed for H5 Highly Pathogenic Avian Influenza in China," *Annals of the New York Academy of Sciences* 1081 (2006): 182–92, doi.org/10.1196/annals.1373.022.

19 Hualan Chen, "Avian Influenza Vaccination: The Experience of China," *Review of Science and Technology of the Office Internationale de Epizootiques*, 28, no. 1 (2009): 268.

20 Chen, "Avian Influenza Vaccination," 268.

21 Chen and Bu, "Development and Application of Avian Influenza Vaccines," 153–62.

22 Tian, Guobin [田国彬], "H5 亚型禽流感灭活疫苗的研制及应用回顾与展望," 445.

23 H. Gu, "Feasibility of Production of Human AI Vaccine in AI Vaccine Manufacturers in China," WHO/OIE/FAO *Consultation on the Feasibility of Human Influenza Vaccine Production in Veterinary Production Facilities*, Geneva, April 27, 2006, www.who.int /influenza/vaccines/GuHong_MoA_China.pdf. By 2006, an official from the Veterinary Bureau claimed that China had a minimum annual capacity of 46 billion vaccine doses, of which 25 billion doses were allotted for domestic use. As the first reassortant vaccine ever commercially produced for H5 avian influenza, export markets soon appeared when the virus spread throughout the world, with sales to Vietnam, Mongolia, and Egypt totaling more than 500 million doses.

24 Chen, "Avian Influenza Vaccination," 270.

25 D. Cyranowski, "China Steps Up Drive to Vaccinate All Domestic Birds," *Nature* 436 (2005): 406, doi.10.1038/438406b. The Chinese policy reads: "对所有的家禽施行强制免疫, 免疫密度应达到100% (有特殊要求不免疫的家禽除外)。" The parenthetical phrase refers to the fact that special "vaccine-free" birds are excepted, for example, sentinel flocks. See Ministry of Agriculture [农业部], 关于印发《加强高致病性禽流感免疫的操作规范》的通知, 农医发 [2005] 27号.

26 Juan Lubroth (FAO) suggested that "the logistics of herding in all those loose chickens and ducks is a little more difficult." Quoted in Cyranowski, "China Steps Up Drive," 406.

27 There are, in fact, five methodologies; however, the remaining three are considered supplemental. "Threatened zones" refers to a continuation of "buffer zone" vaccination around outbreak sites, whereas the policy calls for supplementary immunization on border zones and migratory-bird or wetland zones. I discuss the distinction of *guimo* and *sanyang* farms at greater length in chapter 3.

28 Note, however, that broilers are expected to get only one vaccine because they are typically slaughtered at under fifty days of age.

29 "Every spring, fall institute according to standard protocols one complete mass immunization [*jizhong mianyi*]." Ministry of Agriculture [农业部], 关于印发《加强高致病性禽流感免疫的操作规范》的通 知, 农医发 [2005] 27号.

30 Specifically, the link between free pharmaceutical treatment and mass campaign is the model that led to the control of schistosomiasis. This should be contrasted with another mass campaign model, more famous but actually mostly unsuccessful, that focused on prevention activities such as snail eradication (a vector of schistosomiasis). See Miriam Gross, *Farewell to the God of Plague: Chairman Mao's Campaign to Deworm China* (Oakland: University of California Press, 2016).

31 G. J. D. Smith et al., "Emergence and Predominance of an H5N1 Influenza Variant in China," *Proceedings of the National Academy of Sciences of the United States of America* 103, no. 45 (November 7, 2006): 16936–41. Smith and colleagues specify in a letter responding to critics that their claim is a "speculation."

32 Dennis Normile, "Is China Coming Clean on Bird Flu?," *Science* 314, no. 5801 (2006): 905.

33 Ian Scoones and Paul Forster, "The International Response to Highly Pathogenic Avian Influenza: Science, Policy and Politics" (STEPS Working Paper 10, STEPS Centre, Brighton, 2008).

34 Amy L. S. Staples, *The Birth of Development: How the World Bank, Food and Agriculture Organization, and World Health Organization Changed the World, 1945-1965* (Kent, OH: Kent State University Press, 2006), 100.

35 FAO, *TCP Manual: Managing the Decentralized Technical Cooperation Programme*, June 2009, 5.

36 Staples, *The Birth of Development*, 95.

37 Orr again claimed that the FAO is "above politics" because its objective is "meeting the basic human requirement, food." Compare with the "minimalist biopolitics" of humanitarian organizations, such as Doctors without Borders. See Redfield, *Life in Crisis*.

38 FAO and OIE, "A Global Strategy for the Progressive Control of Highly Pathogenic Avian Influenza (HPAI)," November 2005; FAO Representation in China, "China and FAO: Achievements and Success Stories," 2011, www.fao.org/3/a-at005e.pdf.

39 Quoted in Staples, *Birth of Development*, 84.

40 Staples, *Birth of Development*.

41 Andrew B. Kipnis, *Producing Guanxi: Sentiment, Self, and Subculture in a North China Village* (Durham, NC: Duke University Press, 1997); Judith Farquhar, *Appetites: Food and Sex in Postsocialist China* (Durham, NC: Duke University Press, 2002); Katherine A. Mason, "To Your Health! Toasting, Intoxication and Gendered Critique among Banqueting Women," *China Journal* 69 (2013): 108–33.

42 See Mayfair Yang, *Gifts, Favors, and Banquets: The Art of Social Relationships in China* (Ithaca, NY: Cornell University Press, 1994).

43 On *guanxi* as "local moral world," see Yunxiang Yan, *Private Life under Socialism: Love, Intimacy, and Family Change in a Chinese Village, 1949-1999* (Stanford, CA: Stanford University Press, 2003); see also the discussion in John Osburg, *Anxious Wealth: Money and Morality among China's New Rich* (Stanford, CA: Stanford University Press, 2013).

44 FAO Emergency Center for Transboundary Animal Diseases (ECTAD), Avian Influenza (AI) Project Office, *Avian Influenza Highlights from China*, January 2009. Later reclassified as *China HPAI Highlights* 1 (January 2009).

45 Staples, *Birth of Development*, 100.

46 M. M. J. Fischer, "A Tale of Two Genome Institutes: Qualitative Networks, Charismatic Voice, and R&D Strategies—Juxtaposing GIS Biopolis and BGI," *Science, Technology and Society* 23, no. 2 (2018): 271–88; see also Annelise Riles, *The Network Inside Out* (Ann Arbor: University of Michigan Press, 2000).

47 See, for instance, Hallam Stevens, *Life out of Sequence: A Data-Driven History of Bioinformatics* (Chicago: University of Chicago Press, 2013).

48 See Celia Lowe, *Wild Profusion: Biodiversity Conservation in an Indonesian Archipelago* (Princeton, NJ: Princeton University Press, 2006).

49 Marius Gilbert, "Spatial Modeling of H5N1: A Review," 3rd International Workshop on Community-Based Data Synthesis, Analysis and Modeling of Highly Pathogenic Avian Influenza H5N1 in Asia, Beijing, 2011.

50 Marius Gilbert et al., "Mapping H5N1 Highly Pathogenic Avian Influenza Risk in Southeast Asia," *Proceedings of the National Academy of Sciences of the United States of America* 105, no. 12 (March 25, 2008): 4771, doi.org/10.1073/pnas.0710581105.

51 V. Martin et al., "Spatial Distribution and Risk Factors of Highly Pathogenic Avian Influenza (HPAI) H5N1 in China," *PLoS Pathogens* 7, no. 3 (2011): e1001308.

52 Martin et al., "Spatial Distribution," e1001308.

53 The official was speaking in English, but the phrase *backyard and grazing farms* is a common translation of the Chinese veterinary classification "*sanyang* farms." See my discussion in chapter 3 on "scattered" (*sanyang*) and "scale" (*guimo*) farms.

54 On humanitarian intervention, see Redfield, *Life in Crisis*.

55 Robert Peckham and Ria Sinha, "Satellites and the New War on Infection: Tracking Ebola in West Africa," *Geoforum* 80, Suppl. C (March 1, 2017).

56 P. Fisher, "The Pixel: A Snare and a Delusion," *International Journal of Remote Sensing* 18, no. 3 (February 1, 1997): 679–85, doi.org/10.1080/014311697219015.

57 Peng Li et al., "Changes in Rice Cropping Systems in the Poyang Lake Region, China during 2004–2010," *Journal of Geographical Sciences* 22, no. 4 (2012): 653–68.

58 Francois Jullien, *Detour and Access: Strategies of Meaning in China and Greece* (New York: Zone, 2000).

CHAPTER SIX: OFFICE VETS AND DUCK DOCTORS

1 For example, Mike Davis, *The Monster at Our Door: The Global Threat of Avian Flu* (New York: New Press, 2005).

2 See Lyle Fearnley, "Epidemic Intelligence: Langmuir and the Birth of Disease Surveillance," *Behemoth* 3, no. 3 (2010): 37–56, on the invention of the CDC's Epidemic Intelligence Service and the distinction between "study-section research grant epidemiology" and "epidemic intelligence" or field epidemiology. Note the Chinese translation *xianchang*, a term meaning "on the spot" and used in emergency response, journalism, etc. By contrast, the anthropological "field" is referred to as *tianye*, or "open (uncultivated) fields."

3 Joan Kaufman, "China's Heath Care System and Avian Influenza Preparedness," *Journal of Infectious Diseases* 197, Suppl. 1 (2008): s8.

4 Kaufman, "China's Health Care System," s11.

5 The narrative is developed fully in Pfeiffer's textbook, *Veterinary Epidemiology: An Introduction* (Ames, IA: Wiley-Blackwell, 2010). Following the first module of the training program, an unpublished Chinese translation of the text was passed among the trainees.

6 Katherine Mason, *Infectious Change: Reinventing Chinese Public Health after an Epidemic* (Palo Alto, CA: Stanford University Press, 2016).

7 Although the veterinary system uses the term *yibing* (epidemic disease, plague) rather than *jibing* (disease), and avoids the term *yufang* (prevention) altogether,

the official English translation is Centers for Animal Disease Control, with the acronym CADC.

8   Mason notes that many of her informants considered Robert Fontaine, a U.S. CDC epidemiologist who ran the FETP in Beijing, their "greatest hero." Mason, *Infectious Change*, 12.

9   Mason, *Infectious Change*, 3.

10 Mason, *Infectious Change*, 24.

11 João Guilherme Biehl and Adriana Petryna, *When People Come First: Critical Studies in Global Health* (Princeton, NJ: Princeton University Press, 2013); Vincanne Adams, *Metrics: What Counts in Global Health* (Princeton, NJ: Princeton University Press, 2016).

12 I'm grateful to Michael Fischer for suggesting this term to me.

13 On personal familiarity in post-reform rural medical practice, see Judith Farquhar, "Market Magic: Getting Rich and Getting Personal in Medicine after Mao," *American Ethnologist* 23, no. 2 (1996): 239–57.

14 See Dorothy Porter, "Stratification and Its Discontents: Professionalization and Conflict in the British Public Health Service, 1848–1914," in *A History of Education in Public Health: Health That Mocks the Doctors' Rules*, ed. Elizabeth Fee and Roy Acheson (Oxford: Oxford University Press, 1991).

15 *Dangdai Zhongguo de xumuye* [Contemporary China: animal husbandry], *"Dang dai Zhongguo" cong shu bian ji bu* (Beijing: Dang da Zhongguo chu ban she, 2009), 223.

16 *Dangdai Zhongguo de xumuye*, 224.

17 *Dangdai Zhongguo de xumuye*, 230.

18 China's *fangyi* system is largely based on the antiplague system of the Soviet Union, although it also has roots in the great Manchurian plague epidemics and Wu Liande's establishment of the Manchurian Plague Prevention Service. Perhaps more pertinently, unlike the Soviet system, China established anti-epidemic stations in every commune. On the Soviet antiplague system, see Sonia Ouagrham-Gormley, "Growth of the Anti-Plague System during the Soviet Period," *Critical Reviews in Microbiology* 32, no. 1 (2006): 33–46; on the Manchurian plague, see Christos Lynteris, *Ethnographic Plague: Configuring Disease on the Chinese-Russian Frontier* (London: Palgrave Macmillan, 2016); Xianglin Lei, *Neither Donkey nor Horse: Medicine in the Struggle over China's Modernity* (Chicago: University of Chicago Press, 2014).

19 Yuan Haifeng, "基层畜牧兽医站管理的一些做法和体会." 兽医动态 12, no. 8 (1986): 55–56.

20 Zheng Hong'e et al., "对乡村社会风险管理体系及存在问题的反思: 以禽流感的风险应对为例," *Journal of Nanjing Agricultural University* (Social Sciences Edition) 10, no. 4 (2010): 113–20.

21 Qu Yanchun, "Jitihua shiqi de nongcun gonggongpin gongji: jingyan yu jiejian" (Rural public goods supply of the people's commune period: experiences and lessons), *Anhui nongye kexue* 39, no. 3 (2011): 1782–84.

22 Zheng Hong'e et al., "对乡村社会风险管理体系及存在问题的反思," 114.

23 *Dangdai Zhongguo de xumuye*, 225.

24 See, for example, "发挥"赤脚兽医"的作用," 浙江日报, 1972-11-22, 第二版面; 梁雄, "赤脚兽医," 浙江日报, 1973-04-22, 第四版面; 张正直, 李西富, 赵国际, "一位'傻'赤脚兽医"山东农业1998年12期."

25 Sigrid Schmalzer, *Green Revolution, Red Revolution: Scientific Farming in Socialist China* (Chicago: University of Chicago Press, 2016), 125.

26 Hong'e Zheng et al., "对乡村社会风险管理体系及存在问题的反思," 114.

27 See Laurence A. Schneider, *Biology and Revolution in Twentieth-Century China* (Lanham, MD: Rowman and Littlefield, 2005); Denis Fred Simon and Merle Goldman, *Science and Technology in Post-Mao China* (Cambridge, MA: Council on East Asian Studies, 1989).

28 On the "technocratic" state, see Joel Andreas, *Rise of the Red Engineers: The Cultural Revolution and the Origins of China's New Class* (Stanford, CA: Stanford University Press, 2009); on professionals, see Jingqing Yang, "Professors, Doctors, and Lawyers: The Variable Wealth of the Professional Classes," in *The New Rich in China: Future Rulers, Present Lives*, ed. David Goodman (London: Routledge, 2009).

29 Chengyue Li et al., "An Evaluation of China's New Rural Cooperative Medical System: Achievements and Inadequacies from Policy Goals," BMC *Public Health* 15 (2015): 1079, doi.org/10.1186/s12889-015-2410-1.

30 Yu Huang, "Neoliberalizing Food Safety Control: Training Licensed Fish Veterinarians to Combat Aquaculture Drug Residues in Guangdong," *Modern China* 42, no. 5 (September 2016): 535–65, doi.org/10.1177/0097700415605322.

31 Yuan Haifeng, "基层畜牧兽医站管理的一些做法和体会." 兽医动态, 55–56. Yuan, a county-level administrator, even argues that "for a long time now, the extent of base-level station work has been limited within the constrained, small circle of epidemic prevention and disease eradication," which neither is suitable to supporting the growth of a commercial livestock economy nor offers opportunities for the remuneration of veterinary staff to increase (55).

32 Hong'e Zheng et al., "对乡村社会风险管理体系及存在问题的反思: 以禽流感的风险应对为例," 115.

33 Huang, "Neoliberalizing Food Safety Control."

34 Liang Ruihua, 高致病性禽流感疫病控制模式与宏观管理研究. 武汉: 武汉理工大学博士学位论文, 2007.

35 Cui Yuying, "入世: 解读兽医协会," *Roupin Weisheng* 3, no. 213 (2002): 18.

36 Jiang Kaiyu, "当代兽医体制改革的方向," *Xin Shouyi* 4 (2005).

37 Cui Yuying, "入世," 18.

38 Hong'e Zheng et al., "对乡村社会风险管理体系及存在问题的反思: 以禽流感的风险应对为例," 115.

39 Andrew Kipnis, "*Suzhi*: A Keyword Approach," *China Quarterly* 186, no. 1 (2006): 297; Ann Anagnost, "The Corporeal Politics of Quality (*Suzhi*)," *Public Culture* 16, no. 2 (2004): 189–208; Hairong Yan, "Neoliberal Governmentality and Neohumanism: Organizing *Suzhi*/Value Flow through Labor Recruitment Networks," *Cultural Anthropology* 18, no. 4 (2003).

40 As Ann Anagnost puts it, for example, "The discourse of *suzhi* appears most elaborated in relation to two figures: the body of the rural migrant, which exemplifies

suzhi by its apparent absence, and the body of the middle-class only child, which is fetishized as a site for the accumulation of the very dimensions of *suzhi* wanting in its 'other.'" Anagnost, "The Corporeal Politics of Quality," 190.

41 Mason, *Infectious Change.*

42 "China Field Epidemiology Training Program for Veterinarians (FETPV): Module 1: Introduction to Veterinary Epidemiology, 29 November–22 December," unpublished document, n.d.: principles and concepts of field epidemiology for disease prevention and control; design and assessment of disease surveillance and networks using multilevel and multidisciplinary approaches; planing [*sic*] and conduct of effective outbreak investigations under a variety of field situations.

43 Stephen J. Collier, *Post-Soviet Social: Neoliberalism, Social Modernity, Biopolitics* (Princeton, NJ: Princeton University Press, 2011), 164.

44 Theodore M. Porter, *The Rise of Statistical Thinking: 1820–1900* (Princeton, NJ: Princeton University Press, 2011); Ian Hacking, *The Taming of Chance* (Cambridge: Cambridge University Press, 1990).

45 Jie Yang, *Unknotting the Heart: Unemployment and Therapeutic Governance in China* (Ithaca, NY: Cornell University Press, 2005), 11.

46 Hsuan-Ying Huang, "Untamed Jianghu or Emerging Profession: Diagnosing the Psycho-Boom amid China's Mental Health Legislation," *Culture, Medicine, and Psychiatry* 42, no. 2 (June 2018): 371–400, doi.org/10.1007/s11013-017-9553-8.

47 Yang, *Unknotting the Heart,* 11.

48 Zuoyue Wang, "Science and the State in Modern China," *Isis* 98 (2007): 558–70.

49 Schneider, *Biology and Revolution,* 271.

50 My approach here draws inspiration from Paul Rabinow's grounding of normative questions of science in everyday practices, as well as Steven Shapin's resituating of Mertonian sociology of science in the context of "shop-floor" ethics. See Paul Rabinow, *Making PCR: A Story of Biotechnology* (Chicago: University of Chicago Press, 1996); Steven Shapin, "Who Is the Industrial Scientist? Commentary from Academic Sociology and from the Shop-Floor in the United States, ca. 1900–ca. 1970," in *Knowledge and Social Order: Rethinking the Sociology of Barry Barnes*, ed. Massimo Mazzotti (Aldershot, UK: Ashgate, 2008).

51 "China FETPV 5th Steering Committee Meeting: Minutes," January 18, 2012, Ministry of Agriculture Meeting Room 105.

52 See Hui Li, "The Countryside Vets the State Left Behind," *Sixth Tone*, June 14, 2018, www.sixthtone.com/news/1002439/The%20Countryside%20Vets%20the%20State%20Left%20Behind.

53 See Kipnis, "*Suzhi*," 297–98.

54 Hong'e Zheng et al., "对乡村社会风险," 116.

55 When Hao told a government official about it, though, the official told him "you can never ever bring up avian influenza; only government offices can bring up avian influenza," so Hao began calling it a cure for *ganmaoxing* (a play on *liuxingxing ganmao*, or "influenza").

1 This in no way means that it was the first highly pathogenic avian influenza (HPAI) virus outbreak in the United States. Indeed, the first major reported outbreaks of HPAI took place in Pennsylvania in the 1980s.

2 U.S. Congress, Senate, Committee on Agriculture, Nutrition, and Forestry, *Highly Pathogenic Avian Influenza: The Impact on the U.S. Poultry Sector and Protecting U.S. Poultry Flocks: Hearing Before the Committee on Agriculture, Nutrition, and Forestry*, 114th Cong., 1st Sess., 2015, 17–18.

3 See Sarah Franklin, Celia Lury, and Jackie Stacey, *Global Nature, Global Culture* (London: SAGE, 2000), 75. Franklin, Lury, and Stacey identify erasure of context in fields such as modern genetics, where intrinsic genes are seen as the primary or fundamental unit of life.

4 Nicholas B. King, "Security, Disease, Commerce: Ideologies of Postcolonial Global Health," *Social Studies of Science* 32, nos. 5–6 (2002): 773.

5 Mary Douglas, *Purity and Danger: An Analysis of Concepts of Pollution and Taboo* (New York: Routledge, 2015).

6 Priscilla Wald, *Contagious: Cultures, Carriers, and the Outbreak Narrative* (Durham, NC: Duke University Press, 2008).

7 Quoted in Maryn McKenna, "The Race to Find the Next Pandemic—Before It Finds Us," *Wired*, April 12, 2018, www.wired.com/story/the-race-to-find-the-next-pandemic-before-it-finds-us. Daszak is one of the most ecologically minded of the researchers working on emerging infectious diseases.

8 Warwick Anderson, "Natural Histories of Infectious Disease: Ecological Vision in Twentieth-Century Biomedical Science," *Osiris* 19 (2004): 51–52. More recently, Anderson has suggested that Burnet and other Australian microbiologists learned much from the distinctive "settler colonial" problems of environmental disturbance, pest control, and related ecological management. See Warwick Anderson, "Postcolonial Ecologies of Parasite and Host: Making Parasitism Cosmopolitan," *Journal of the History of Biology* 49, no. 2 (2016): 241–59; Warwick Anderson, "Nowhere to Run, Rabbit: The Cold-War Calculus of Disease Ecology," *History and Philosophy of the Life Sciences* 39, no. 2 (June 2017): 13, doi.org/10.1007/s40656-017-0140-7.

9 Karin Knorr-Cetina, *Epistemic Cultures: How the Sciences Make Knowledge* (Cambridge, MA: Harvard University Press, 2009).

10 Gary Lee Downey and Joseph Dumit, *Cyborgs and Citadels: Anthropological Interventions in Emerging Sciences and Technologies* (Santa Fe, NM: School of American Research Press, 1997).

11 Helga Nowotny, Peter Scott, and Michael Gibbons, *Re-thinking Science: Knowledge and the Public in an Age of Uncertainty* (Cambridge: Polity, 2001).

12 Michel Callon and Vololona Rabeharisoa, "Research 'in the Wild' and the Shaping of New Social Identities," *Technology in Society* 25, no. 2 (2003): 193–204; Brian Wynne, "May the Sheep Safely Graze? A Reflexive View of the Expert–Lay Knowledge Divide," in *Risk, Environment, and Modernity: Towards a New Ecology*, ed. Scott Lash, Bronislaw Szerszynski, and Brian Wynne (London: SAGE, 1996); Marilyn

Strathern, "Re-describing Society," *Minerva* 41, no. 3 (2003): 263–76, doi.org/10.1023/A:1025586327342; Paul Rabinow and Gaymon Bennett, *Designing Human Practices: An Experiment with Synthetic Biology* (Chicago: University of Chicago Press, 2012); Anthony Stavrianakis, "From Anthropologist to Actant (and Back to Anthropology): Position, Impasse, and Observation in Sociotechnical Collaboration," *Cultural Anthropology* 30, no. 1 (February 17, 2015): 169–89, doi.org/10.14506/ca30.1.09.

13 Ulrich Beck, *Risk Society: Towards a New Modernity* (London: SAGE, 1992).

14 A brief sample includes Kim Fortun, "Ethnography in Late Industrialism," *Cultural Anthropology* 27, no. 3 (November 12, 2012): 446–64; Janelle Lamoreaux, "What if the Environment Is a Person? Lineages of Epigenetic Science in a Toxic China," *Cultural Anthropology* 31, no. 2 (May 4, 2016): 188–214, doi.org/10.14506/ca31.2.03; Hannah Landecker, "Antibiotic Resistance and the Biology of History," *Body and Society* 22, no. 4 (2016): 19–52; Stefan Helmreich, *Alien Ocean: Anthropological Voyages in Microbial Seas* (Berkeley: University of California Press, 2009).

15 Carlo Caduff, "After the Next: Notes on Serial Novelty," *Medicine Anthropology Theory* 5, no. 4 (September 10, 2018): 86–105, doi.org/10.17157/mat.5.4.623.

16 A. M. Kleinman et al., "Asian Flus in Ethnographic and Political Context: A Biosocial Approach," *Anthropology and Medicine* 15, no. 1 (2008): 1, dx.doi.org/10.1080/13648470801918968.

17 Wald, *Contagious*, 8.

18 Madhur S. Dhingra et al., "Global Mapping of Highly Pathogenic Avian Influenza H5N1 and H5Nx Clade 2.3.4.4 Viruses with Spatial Cross-Validation," *ELife* 5 (November 25, 2016): e19571, doi.org/10.7554/eLife.19571.

19 Alexandre H. Hirzel and Gwenaëlle Le Lay, "Habitat Suitability Modelling and Niche Theory," *Journal of Applied Ecology* 45, no. 5 (October 1, 2008): 1372–81, doi.10.1111/j.1365-2664.2008.01524.x.

20 Shauna-Lee Chai et al., "Using Risk Assessment and Habitat Suitability Models to Prioritise Invasive Species for Management in a Changing Climate," *PLoS ONE* 11, no. 10 (October 21, 2016), doi.org/10.1371/journal.pone.0165292.

21 Ludwik Fleck, *Genesis and Development of a Scientific Fact*, ed. Thaddeus J. Trann and Robert T. Merton (Chicago: University of Chicago Press, 1981), 23.

22 Fleck, *Genesis and Development of a Scientific Fact*, 51.

23 On the tropics and tropical medicine, see *Warm Climates and Western Medicine: The Emergence of Tropical Medicine, 1500–1900*, ed. David Arnold (Amsterdam: Editions Rodopi, 1996); Warwick Anderson, *Colonial Pathologies: American Tropical Medicine, Race, and Hygiene in the Philippines* (Durham, NC: Duke University Press, 2006); Aihwa Ong, *Fungible Life: Experiment in the Asian City of Life* (Durham, NC: Duke University Press, 2016).

24 Ulrich Beck, *World Risk Society* (Cambridge: Polity, 2009); Anna Lowenhaupt Tsing, *The Mushroom at the End of the World: On the Possibility of Life in Capitalist Ruins* (Princeton, NJ: Princeton University Press, 2015); Bruno Latour, *Facing Gaia: Eight Lectures on the New Climatic Regime*, trans. Catherine Porter (Cambridge: Polity, 2018).

25 Tim Ingold, *The Perception of the Environment: Essays on Livelihood, Dwelling, and Skill* (London: Routledge, 2000), 209.

26 Tim Ingold, *Lines: A Brief History* (London: Routledge, 2016), 92.

27 Tim Ingold, *Being Alive: Essays on Movement, Knowledge and Description* (London: Routledge, 2011), 153. Ingold's language is clearly adopted from STS texts on scientific practice, such as Bruno Latour's essay "Circulating Reference." At times, Ingold does acknowledge that this is more of the "'official' view of what is supposed to happen" than how science is actually practiced (154).

28 Latour, *Facing Gaia*, 130.

29 Jeremy Vetter, "Introduction," and James Rodger Fleming, "Planetary-Scale Fieldwork: Harry Wexler on the Possibilities of Ozone Depletion and Climate Control," in *Knowing Global Environments: New Historical Perspectives on the Field Sciences*, ed. Jeremy Vetter (New Brunswick, NJ: Rutgers University Press, 2010).

30 Dennis Carroll et al., "The Global Virome Project," *Science* 359, no. 6378 (February 20, 2018): 872.

31 Carroll et al., "The Global Virome Project," 872.

32 Dennis Carroll et al., "Building a Global Atlas of Zoonotic Viruses," *Bulletin of the World Health Organization* 96, no. 4 (April 1, 2018): 292, doi.org/10.2471/BLT.17 .205005.

33 Carroll et al., "The Global Virome Project," 872. See also Melinda Cooper, "Preempting Emergence: The Biological Turn in the War on Terror," *Theory, Culture and Society* 23, no. 4 (July 2006): 113–35, doi.org/10.1177/0263276406065121.

34 Xiaodong Wang and Jua Shan, "China to Help ID Unknown Lethal Viruses," *China Daily*, May 22, 2018.

35 Franklin, Lury, and Stacey, *Global Nature, Global Culture*; Tobias Rees, "Humanity/Plan; or, On the 'Stateless' Today (Also Being an Anthropology of Global Health)," *Cultural Anthropology* 29, no. 3 (August 11, 2014): 457–78, doi.org/10.14506/ca29.3 .02; Hallam Stevens, "Globalizing Genomics: The Origins of the International Nucleotide Sequence Database Collaboration," *Journal of the History of Biology* 51, no. 4 (2018): 657–91; Michael M. J. Fischer, "Biopolis: Asian Science in the Global Circuitry," *Science, Technology and Society* 18, no. 3 (2013): 381–406.

36 Carroll et al., "The Global Virome Project," 873. See also Amy Hinterberger and Natalie Porter, "Genomic and Viral Sovereignty: Tethering the Materials of Global Biomedicine," *Public Culture* 27, no. 2–76 (May 1, 2015): 361–86, doi.org/10.1215 /08992363-2841904.

37 Stephen J. Collier, "Global Assemblages," *Theory, Culture and Society* 23 (2006): 400.

38 Stephen J. Collier and Aihwa Ong, "Global Assemblages, Anthropological Problems," in *Global Assemblages: Technology, Politics, and Ethics as Anthropological Problems*, ed. Aihwa Ong and Stephen J. Collier (Malden, MA: Blackwell, 2005); Bruno Latour, *Science in Action: How to Follow Scientists and Engineers through Society* (Cambridge, MA: Harvard University Press, 1987); John Law, "On the Methods of Long Distance Control: Vessels, Navigation, and the Portuguese Route to India," in *Power, Action, and Belief: A New Sociology of Knowledge?*, ed. John Law (London: Routledge, 1987). On the *qualitative* networks in global science, see M. M. J. Fischer, "A

Tale of Two Genome Institutes: Qualitative Networks, Charismatic Voice, and R&D Strategies—Juxtaposing GIS Biopolis and BGI," *Science, Technology and Society* 23, no. 2 (2018): 271–88.

39 Marius Gilbert, Xiangming Xiao, and Timothy P. Robinson, "Intensifying Poultry Production Systems and the Emergence of Avian Influenza in China: A 'One Health/Ecohealth' Epitome," *Archives of Public Health* 75, no. 48 (2017): 1–7. See also FAO, *World Livestock 2013—Changing Disease Landscapes*, Rome, 2013.

40 Gilbert, Xiao, and Robinson, "Intensifying Poultry Production Systems," 6–7.

41 James H. Dunk et al., "Human Health on an Ailing Planet—Historical Perspectives on Our Future," ed. Debra Malina, *New England Journal of Medicine* 381, no. 8 (August 22, 2019): 778–82, doi.org/10.1056/NEJMms1907455.

42 Richard Horton and Selina Lo, "Planetary Health: A New Science for Exceptional Action," *Lancet* 386, no. 10007 (November 2015): 1921–22, doi.org/10.1016/S0140-6736(15)61038-8.

43 Peter Sloterdijk, *In the World Interior of Capital: Towards a Philosophical Theory of Globalization*, trans. Wieland Hoban (Cambridge: Polity, 2015); Donna Haraway, *Staying with the Trouble: Making Kin in the Cthulucene* (Durham, NC: Duke University Press, 2016); Latour, *Facing Gaia*.

44 Marisa Wilson et al., "Why 'Culture' Matters for Planetary Health," *Lancet Planetary Health* 2, no. 11 (November 2018): e467–68, doi.org/10.1016/S2542-5196(18)30205-5; Meike Wolf, "Is There Really Such a Thing as 'One Health'? Thinking about a More Than Human World from the Perspective of Cultural Anthropology," *Social Science and Medicine* 129 (March 2015): 5–11, doi.org/10.1016/j.socscimed.2014.06.018.

45 Ziping Wu, "Antibiotic Use and Antibiotic Resistance in Food-Producing Animals in China" (OECD Food, Agriculture and Fisheries Papers no. 134, OECD Publishing, Paris, 2019, dx.doi.org/10.1787/4ADba8c1-en).

46 Gilles Deleuze and Félix Guattari, "10,000 BC: The Geology of Morals (Who Does the Earth Think It Is?)," in *A Thousand Plateaus*, trans. Brian Massumi (Minneapolis: University of Minnesota Press, 1987). See also Manuel Delanda, *A New Philosophy of Society: Assemblage Theory and Social Complexity* (London: Continuum, 2006).

47 Bruno Latour, "Give Me a Laboratory and I Will Raise the World," in *Science Observed: Perspectives on the Social Study of Science*, ed. Karin D. Knorr-Cetina and Michael Mulkay (London: SAGE, 1983).

POSTSCRIPT

1 On February 11, 2020, the International Committee on Taxonomy of Viruses officially named the virus 'Severe Acute Respiratory Syndrome coronavirus 2' (SARS-CoV-2) and named the disease caused by this virus 'coronavirus disease 2019' (COVID-19). Like WHO, to avoid confusion with SARS, I will refer to the virus as covid-19 virus.

2 For example, Chris Buckley, "Losing Track of Time in the Epicenter of China's Coronavirus Outbreak," *New York Times*, February 5, 2020.

3   Peng Zhou et al., "A Pneumonia Outbreak Associated with a New Coronavirus of Probable Bat Origin," *Nature*, February 3, 2020, doi.org/10.1038/s41586-020-2012-7.

4   "Wuhan Virus: Rats and Live Wolf Pups on the Menu at China Food Market Linked to Virus Outbreak," *Straits Times*, January 22, 2020.

5   Steven Lee Meyers, "China's Omnivorous Markets Are in the Eye of a Lethal Outbreak Once Again," *New York Times*, January 25, 2020.

6   Mei Zhan, "Civet Cats, Fried Grasshoppers, and David Beckham's Pajamas: Unruly Bodies after SARS," *American Anthropologist* 107, no. 1 (2005): 31–42, doi.org/10.1525/aa.2005.107.1.031.

7   Hinterberger and Porter, "Genomic and Viral Sovereignty."

8   See, for example, Ben Hu et al., "Discovery of a Rich Gene Pool of Bat SARS-Related Coronaviruses Provides New Insights into the Origin of SARS Coronavirus," *PLoS Pathogens* 13, no. 11 (November 30, 2017): e1006698, doi.org/10.1371/journal.ppat.1006698.

9   Fischer, "Biopolis," 384.

10 Gilbert, Xiao, and Robinson, "Intensifying Poultry Production Systems," 6–7.

11 Yin Li et al., "Closure of Live Bird Markets Leads to the Spread of H7N9 Influenza in China," *PLoS ONE* 13, no. 12 (December 12, 2018): e0208884, doi.org/10.1371/journal.pone.0208884.

12 Mimi Lau, "Black Market for Live Chickens Thrives in China Despite Bird Flu Bans," *South China Morning Post*, February 27, 2017.

PRIMARY SOURCES AND ARCHIVES

Beveridge, W. I. B. "Where Did Red Flu Come From?" *New Scientist* (March 23, 1978): 790.

Buckley, Chris. "Losing Track of Time in the Epicenter of China's Coronavirus Outbreak." *New York Times*, February 5, 2020.

Cappelle, Julien, Zhao Delong, Marius Gilbert, Martha I. Nelson, Scott H. Newman, John Y. Takekawa, Nicolas Gaidet, et al. "Risks of Avian Influenza Transmission in Areas of Intensive Free-Ranging Duck Production with Wild Waterfowl." *EcoHealth* 11 (2014): 109–19. doi:10.1007/s10393-014-0914-2.

Carroll, Dennis, Peter Daszak, Nathan D. Wolfe, George F. Gao, Carlos M. Morel, Subhash Morzaria, Ariel Pablos-Méndez, Oyewale Tomori, and Jonna A. K. Mazet. "The Global Virome Project." *Science* 359, no. 6378 (February 20, 2018): 872–74.

Carroll, Dennis, Brooke Watson, Eri Togami, Peter Daszak, Jonna A. K. Mazet, Cara J. Chrisman, Edward M. Rubin, et al. "Building a Global Atlas of Zoonotic Viruses." *Bulletin of the World Health Organization* 96, no. 4 (April 1, 2018): 292–94. doi.org/10.2471/BLT.17.205005.

Chai, Shauna-Lee, Jian Zhang, Amy Nixon, and Scott Nielsen. "Using Risk Assessment and Habitat Suitability Models to Prioritise Invasive Species for Management in a Changing Climate." *PLoS ONE* 11, no. 10 (October 21, 2016). doi.org/10.1371/journal.pone.0165292.

Chen, Hualan. "Avian Influenza Vaccination: The Experience of China." *Review of Science and Technology of the Office Internationale de Epizooties* 28, no. 1 (2009): 267–74.

Chen, Hualan, and Zhigao Bu. "Development and Application of Avian Influenza Vaccines in China." In *Vaccines for Pandemic Influenza*, edited by Richard W. Compans and Walter A. Orenstein, 153–62. Berlin: Springer Berlin Heidelberg, 2009. doi.org/10.1007/978-3-540-92165-3_7.

Chen, H., G. J. D. Smith, K. S. Li, J. Wang, X. H. Fan, J. M. Rayner, D. Vijaykrishna, et al. "Establishment of Multiple Sublineages of H5N1 Influenza Virus in Asia: Implications for Pandemic Control." *Proceedings of the National Academy of Sciences of the United States of America* 103, no. 8 (2006): 2845–50.

Chen, Tan. "养殖大眼, 前景广阔." 农村农业农民 no. 10 (1999): 21.

"China FETPV 5th Steering Committee Meeting: Minutes." Ministry of Agriculture Meeting Room 105, January 18, 2012.

"China Is Committed to the Fight against H5N1 Avian Influenza, but Challenges Remain; Opening Remarks by Dr. Shigeru Omi, WHO Regional Director for the Western Pacific Region News Conference on China's Response to Avian Influenza." *PR Newswire*, December 23, 2005. www.prnewswire.com/news-releases /china-is-committed-to-the-fight-against-h5n1-avian-influenza-but-challenges -remain-55655887.html.

Chu, Chi-Ming. "The Etiology and Epidemiology of Influenza: An Analysis of the 1957 Epidemic." *Journal of Hygiene, Epidemiology, Microbiology, and Immunology* 2, no. 1 (1958): 1–8.

Chu, C. M., C. H. Andrewes, and A. W. Gledhil. "Influenza in 1948–1949." *Bulletin of the World Health Organization* 3 (1950): 208–9.

Cockburn, Charles W., P. J. Delon, and W. Ferreira. "Origin and Progress of the 1968–69 Hong Kong Influenza Epidemic." *Bulletin of the World Health Organization* 41, nos. 3-4-5 (1969): 345–46.

Cui Yuying [崔玉英]. "入世: 解读兽医协会" 肉品卫生 3, no. 213 (2002): 18–19.

Cyranowski, D. "China Steps Up Drive to Vaccinate All Domestic Birds." *Nature* 436 (2005): 406. doi:10.1038/438406b.

Dangdai Zhongguo de xumuye [Contemporary China: animal husbandry], *"Dangdai Zhongguo" cong shu bian ji bu*. Beijing: Dang da Zhongguo chu ban she, 2009.

Davies, Pete. "The Plague in Waiting." *Guardian*, August 7, 1999.

de Jong, J. C., E. C. Claas, A. D. Osterhaus, R. G. Webster, and W. L. Lim. "A Pandemic Warning?" *Nature* 389, no. 6651 (October 1997): 554. doi.org/10.1038/39218.

Delgado, Christopher L. *Livestock to 2020: The Next Food Revolution*. Rome: Food and Agriculture Organization of the United Nations, 1999.

Dhingra, Madhur S., Jean Artois, Timothy P. Robinson, Catherine Linard, Celia Chaiban, Ioannis Xenarios, Robin Engler, et al. "Global Mapping of Highly Pathogenic Avian Influenza H5N1 and H5Nx Clade 2.3.4.4 Viruses with Spatial Cross-Validation." *ELife* 5 (November 25, 2016): e19571. doi.org/10.7554/eLife.19571.

Ding, Changqing. "Farmed Wild Ducks and Geese: The Missing Link in Understanding the Diversity and Epidemiology of Pathogens Transmitted among Poultry, Wildlife, and People." The 3rd International Workshop on Community-based Data Synthesis, Analysis and Modeling of Highly Pathogenic Avian Influenza H5N1 in Asia. Beijing, November 14–15, 2011.

FAO. "Avian Influenza: Stop the Risk for Humans and Animals at Source." Donor appeal, circa 2005. Accessed April 1, 2009. www.fao.org/avianflu/documents /donor.pdf.

FAO. "Avian Influenza Control and Eradication: FAO's Proposal for a Global Programme." Rome, March 2006.

FAO. "Biosecurity for Highly Pathogenic Avian Influenza: Issues and Options." Rome, 2008.

FAO. *Lessons from HPAI: A Technical Stockpiling of Outputs, Outcomes, Best Practices, and Lessons Learned from the Fights against Highly Pathogenic Avian Influenza in Asia 2005–2011*, FAO Animal Production and Health Paper no. 176. Rome, 2013.

FAO. "Recommendations on the Prevention, Control and Eradication of Highly Pathogenic Avian Influenza (HPAI) in Asia (Proposed with the Support of OIE)." Rome, September 2004.

FAO. *TCP Manual: Managing the Decentralized Technical Cooperation Programme.* June 2009, 5.

FAO. *World Livestock 2013—Changing Disease Landscapes.* Rome, 2013.

FAO and OIE. "A Global Strategy for the Progressive Control of Highly Pathogenic Avian Influenza (HPAI)." November 2005. www.fao.org/avianflu/documents /hpaiglobalstrategy31oct05.pdf.

FAO/WHO/OIE. "Unprecedented Spread of Avian Influenza Requires Broad Collaboration." January 27, 2004. www.fao.org/newsroom/en/news/2004/35988/index .html.

FAO Emergency Center for Transboundary Animal Diseases (ECTAD), Avian Influenza (AI) Project Office. *Avian Influenza Highlights from China* (January 2009) (later reclassified as *China HPAI Highlights* 1 [January 2009]).

FAO Representation in China. "China and FAO: Achievements and Success Stories." 2011. www.fao.org/3/a-at005e.pdf.

Fenner, Frank. "Frank Macfarlane Burnet 1899–1985." *Historical Records of Australian Science* 7, no. 1 (1987).

Finkler, Ditmar. "Influenza." In *Twentieth Century Practice: An International Encyclopedia of Modern Medical Science by Leading Authorities of Europe and America*, vol. 15, edited by Thomas L. Stedman. New York: Publisher's Printing, 1898.

Gargan, Edward. "As Avian Flu Spreads, China Is Seen as Its Epicenter." *New York Times*, December 21, 1997.

Gilbert, Marius. "Spatial Epidemiology of Highly Pathogenic Avian Influenza (HPAI) in Thailand: Exploring the Use of Remote Sensing for Risk Assessment." *Final Report of a Letter of Agreement between the Food and Agriculture Organization of the United Nations and the Université Libre de Bruxelles.* November 2005.

Gilbert, Marius. "Spatial Modeling of H5N1: A Review." The 3rd International Workshop on Community-Based Data Synthesis, Analysis and Modeling of Highly Pathogenic Avian Influenza H5N1 in Asia. Beijing, November 2011.

Gilbert, Marius, P. Chaitaweesup, Tippawon Parakamawongsa, Sith Premashthira, Thanawat Tiensin, Wantanee Kalpravidh, Hans Wagner, and Jan Slingenbergh. "Free-Grazing Ducks and Highly Pathogenic Avian Influenza, Thailand." *Emerging Infectious Diseases* 12, no. 2 (2006): 227–34.

Gilbert, Marius, Xiangming Xiao, Dirk U. Pfeiffer, M. Epprecht, Stephen Boles, Christina Czarnecki, Prasit Chaitaweesub, et al. "Mapping H5N1 Highly Pathogenic Avian Influenza Risk in Southeast Asia." *Proceedings of the National Academy of Sciences of the United States of America* 105, no. 12 (2008): 4769–74. doi.org/10.1073 /pnas.0710581105.

Gilbert, Marius, Xiangming Xiao, and Timothy P. Robinson. "Intensifying Poultry Production Systems and the Emergence of Avian Influenza in China: A 'One Health/Ecohealth' Epitome." *Archives of Public Health* 75, no. 48 (2017): 1–7.

Gu, H. "Feasibility of Production of Human AI Vaccine in AI Vaccine Manufacturers in China." *WHO/OIE/FAO Consultation on the Feasibility of Human Influenza Vaccine Production in Veterinary Production Facilities.* Geneva, April 27, 2006. www.who.int /influenza/vaccines/GuHong_MoA_China.pdf.

Guo, Yuanji, Min Wang, Ping Wing, and Jiming Zhu. "Influenza Ecology in China." In *The Origin of Pandemic Influenza Viruses: Proceedings of the International Workshop on the Molecular Biology and Ecology of Influenza Virus Held in Peking, China, November 10-12, 1982*, edited by W. G. Laver. New York: Elsevier, 1983.

Hoffmann, Erich, Scott Krauss, Daniel Perez, Richard Webby, and Robert G. Webster. "Eight-Plasmid System for Rapid Generation of Influenza Virus Vaccines." *Vaccine* 20, nos. 25-26 (August 19, 2002): 3165-70.

Horton, Richard, and Selina Lo. "Planetary Health: A New Science for Exceptional Action." *Lancet* 386, no. 10007 (November 2015): 1921-22. doi.org/10.1016/S0140 -6736(15)61038-8.

Hu, Ben, Lei-Ping Zeng, Xing-Lou Yang, Xing-Yi Ge, Wei Zhang, Bei Li, Jia-Zheng Xie, et al. "Discovery of a Rich Gene Pool of Bat SARS-Related Coronaviruses Provides New Insights into the Origin of SARS Coronavirus." *PLoS Pathogens* 13, no. 11 (November 30, 2017): e1006698. doi.org/10.1371/journal.ppat .1006698.

Jiang, Kaiyu [汪开毓], "当代兽医体制改革的方向" 新兽医 4 (2005).

Kaplan, M. M. "Relationships between Animal and Human Influenza." *Bulletin of the World Health Organization* 41, nos. 3-4-5 (1969): 2393b.

Kaplan, M. M. "The Role of the World Health Organization in the Study of Influenza." *Philosophical Transactions of the Royal Society of London. Series B, Biological Sciences* 288, no. 1029 (1980): 417-21.

Kaplan, M. M., and W. I. B. Beveridge. "WHO Coordinated Research on the Role of Animals in Influenza Epidemiology." *Bulletin of WHO* 47, no. 4 (1972): 439-42.

Kaplan, M. M., and A. M.-M. Payne. "Serological Survey in Animals for Type A Influenza in Relation to the 1957 Pandemic." *Bulletin of the World Health Organization* 20 (1959): 465-88.

Kilbourne, Edwin. "Recombination of Influenza A Viruses of Human and Animal Origin." *Science* 160, no. 3823 (1968): 74-76.

Kitler, M. E., P. Gavinio, and D. Lavanchy. "Influenza and the Work of the World Health Organization." *Vaccine* 20, Suppl. 2 (2002): S5-S15.

Lau, Mimi. "Black Market for Live Chickens Thrives in China Despite Bird Flu Bans." *South China Morning Post*, February 27, 2017.

Laver, Graeme. "Influenza Virus Surface Glycoproteins, Haemagglutinin, and Neuraminidase: A Personal Account." *Perspectives in Medical Virology* 7 (2002): 31-47.

Li, Peng, Zhiming Feng, Luguang Jiang, Yujie Liu, and Xiangming Xiao. "Changes in Rice Cropping Systems in the Poyang Lake Region, China during 2004-2010." *Journal of Geographical Sciences* 22, no. 4 (2012): 653-68.

Li, Yin, Youming Wang, Chaojian Shen, Jianlong Huang, Jingli Kang, Baoxu Huang, Fusheng Guo, and John Edwards. "Closure of Live Bird Markets Leads to the Spread of H7N9 Influenza in China." *PLoS ONE* 13, no. 12 (December 12, 2018): e0208884. doi.org/10.1371/journal.pone.0208884.

Li, Yuming. "野生动物与特重养殖. 农民日报." March 15, 2001.

Lim K. A., A. Smith, J. H. Hale, and J. Glass. "Influenza Outbreak in Singapore." *Lancet* 273 (1957): 791-96.

Macfarlane Burnet, Frank. *Natural History of Infectious Disease*. 2nd ed. Cambridge: Cambridge University Press, 1953.

"Marius Gilbert." *Prince Mahidol Award Conference 2013: A World United against Infectious Diseases: Cross-sectoral Solutions*. Bangkok, January 28, 2013. www.pmaconference.mahidol.ac.th/index.php?option=com_content&view=article&id=602&Itemid=195.

Martin V., X. Zhou, F. Guo, D. U. Pfeiffer, X. Xiao, D. J. Prosser, and M. Gilbert. "Spatial Distribution and Risk Factors of Highly Pathogenic Avian Influenza (HPAI) H5N1 in China." *PLoS Pathogens* 7, no. 3 (2011).

Meyers, Steven Lee. "China's Omnivorous Markets Are in the Eye of a Lethal Outbreak Once Again." *New York Times*, January 25, 2020.

Ministry of Agriculture [农业部]. 关于印发《加强高致病性禽流感免疫的操作规范》的通知, 农医发 [2005] 27.

"The 1957 Influenza Epidemic." *Indian Journal of Pediatrics*, 24, no. 117 (1957): 354-55.

Payne, A. M. "The Influenza Programme of WHO." *Bulletin of the World Health Organization* 8 (1953): 755-74.

Payne, A. M. "Some Aspects of the Epidemiology of the 1957 Influenza Pandemic." *Proceedings of the Royal Society of Medicine* 51, no. 1009 (1958): 29-38.

Peiris, J. S., W. C. Yu, C. W. Leung, C. Y. Cheung, W. F. Ng, J. M. Nicholls, T. K. Ng, et al. "Re-emergence of Fatal Human Influenza A Subtype H5N1 Disease." *Lancet* 363, no. 9409 (2004): 617-19.

Pfeiffer, Dirk. *Veterinary Epidemiology: An Introduction*. Ames, IA: Wiley-Blackwell, 2010.

Pleschka, S., R. Jaskunas, O. G. Engelhardt, T. Zürcher, P. Palese, and A. García-Sastre. "A Plasmid-Based Reverse Genetics System for Influenza A Virus." *Journal of Virology* 70, no. 6 (June 1996): 4188-92.

Qiao, Chuanling, Guobin Tian, Yongping Jiang, Yanbing Li, Jianzhong Shi, Kangzhen Yu, and Hualan Chen. "Vaccines Developed for H5 Highly Pathogenic Avian Influenza in China." *Annals of the New York Academy of Sciences* 1081 (2006): 182-92. doi.org/10.1196/annals.1373.022.

Shim, Eunha, and Alison P. Galvani. "Evolutionary Repercussions of Avian Culling on Host Resistance and Influenza Virulence." *PLoS ONE* 4, no. 5 (May 11, 2009): e5503.

Shortridge, Kennedy F. "Avian Influenza A Viruses of Southern China and Hong Kong: Ecological Aspects and Implications for Man." *Bulletin of the World Health Organization* 60, no. 1 (1982): 129-35.

Shortridge, Kennedy F. "Pandemic Influenza: Application of Epidemiology and Ecology in the Region of Southern China to Prospective Studies." In *The Origin of Pandemic Influenza Viruses: Proceedings of the International Workshop on the Molecular Biology and Ecology of Influenza Virus Held in Peking, China, November 10-12, 1982*, edited by W. G. Laver. New York: Elsevier, 1983.

Shortridge, K. F., D. J. Alexander, and M. S. Collins. "Isolation and Properties of Viruses from Poultry in Hong Kong Which Represent a New (Sixth) Distinct Group of Avian Paroxymoviruses." *Journal of General Virology* 49 (1980): 255-62.

Shortridge, K. F., W. K. Butterfield, R. G. Webster, and C. H. Campbell. "Isolation and Characterization of Influenza A Viruses from Avian Species in Hong Kong." *Bulletin of the World Health Organization* 55 (1977): 15–20.

Shortridge, Kennedy, and C. H. Stuart-Harris. "An Influenza Epicentre?" *Lancet* 2, no. 8302 (1982): 812–13.

Sichuan Nanyang Special Type Economic Animal and Plant Association. "新型特种养殖: 大雁. 农村新技术," no. 10 (1999).

Sims, L. D., T. M. Ellis, K. K. Liu, K. Dyrting, H. Wong, M. Peiris, Y. Guan, and K. F. Shortridge. "Avian Influenza in Hong Kong 1997–2002." *Avian Diseases* 47, no. 3 (2003): 832–38.

Slingenbergh, J., M. Gilbert, K. de Balogh, and W. Wint. "Ecological Sources of Zoonotic Diseases." *Review of Science and Technology of the Office Internationale de Epizootiques* 23, no. 2 (2004): 469.

Smith, G. J. D., X. H. Fan, J. Wang, K. S. Li, K. Qin, J. X. Zhang, D. Vijaykrishna, et al. "Emergence and Predominance of an H5N1 Influenza Variant in China." *Proceedings of the National Academy of Sciences of the United States of America* 103, no. 45 (November 7, 2006): 16936–41. doi.org/10.1073/pnas.0608157103.

Studies on Poyang Lake Editorial Committee [《鄱阳湖研究》编委会]. 鄱阳湖研究. 上海: 上海科学技术出版社: 新华书店上海发行所发行, 1988.

Thieme, Olaf, and Emanuelle Guerne Bleich. "Poultry Sector Restructuring for Disease Control: Initial Thoughts." FAO Animal Production Service. Rome, April 2007. www.fao.org/docs/eims/upload/239034/ai291e.pdf.

Tian, Guobin [田国彬]. "H5 亚型禽流感灭活疫苗的研制及应用回顾与展望." 《中国农业科学》 40, no. 1 (2007): 444–48.

U.S. Congress. Senate. Committee on Agriculture, Nutrition, and Forestry. *Highly Pathogenic Avian Influenza: The Impact on the U.S. Poultry Sector and Protecting U.S. Poultry Flocks: Hearing Before the Committee on Agriculture, Nutrition, and Forestry.* 114th Cong., 1st Sess., July 7, 2015.

Van Boeckel, Thomas P., Diann Prosser, Gianluca Franceschini, Chandra Biradar, William Wint, Tim Robinson, and Marius Gilbert. "Modelling the Distribution of Domestic Ducks in Monsoon Asia." *Agriculture, Ecosystems and Environment* 141, nos. 3–4 (May 1, 2011): 373–80. doi.org/10.1016/j.agee.2011.04.013.

Wallace, Robert G., HoangMinh HoDac, Richard H. Lathrop, and Walter M. Fitch. "A Statistical Phylogeography of Influenza A H5N1." *Proceedings of the National Academy of Sciences of the United States of America* 104, no. 11 (2007): 4473–78.

Wang, Xiaodong, and Jua Shan. "China to Help ID Unknown Lethal Viruses." *China Daily*, May 22, 2018.

Wang, Yong, Zhiben Jiang, Zhenyu Jin, Hua Tan, and Bing Xu. "Risk Factors for Infectious Diseases in Backyard Poultry Farms in the Poyang Lake Area, China." *PLoS ONE* 8, no. 6 (2013).

Webster, Robert. "Antigenic Variation in Influenza Viruses, with Special Reference to Hong Kong Influenza." *Bulletin of the World Health Organization* 41, nos. 3–4–5 (1969): 483–85.

Webster, R. G., W. J. Bean, O. T. Gorman, T. M. Chambers, and Y. Kawaoka. "Evolution and Ecology of Influenza A Viruses." *Microbiological Reviews* 56, no. 1 (1992): 152–79.

Webster, R. G., and C. H. Campbell. "Studies on the Origin of Pandemic Influenza: IV. Selection and Transmission of 'New' Influenza Viruses in Vivo." *Virology* 62, no. 2 (1974): 404–13.

Webster, R. G., C. H. Campbell, and A. Granoff. "The '*In Vivo*' Production of 'New' Influenza Viruses: III. Isolation of Recombinant Influenza Viruses under Simulated Conditions of Natural Transmission." *Virology* 51, no. 1 (1973): 149–62.

WHO. "General Discussion—Session II." *Bulletin of the World Health Organization* 41, nos. 3-4-5 (1969): 492.

WHO. "Influenza: Introduction." *Bulletin of the World Health Organization* 20, nos. 2–3 (1959): 183–85.

WHO. Official Records of the World Health Organization, "Minutes of the Fourth Session of the Interim Commission," 1947.

WHO. *Weekly Epidemiological Record* 19 (May 10, 1957): 241.

WHO. *Weekly Epidemiological Record* 22 (May 29, 1957): 277.

WHO. *Weekly Epidemiological Record* 23–52 (1957) and 1–13 (1958).

WHO Archives. ARC010-3, Centralized Files, 3rd Generation, Sub-fonds 3, Box 12-133-1. Webster to John Skehal, October 6, 1981.

WHO Archives. ARC010-3, Centralized Files, 3rd Generation, Sub-fonds 3, Box 12-133-1. "Training Courses and Workshops on Influenza—General." Handwritten letter Alan Kendal to Fakhry Assaad, November 5, 1981.

WHO Archives. ARC010-3, Centralized Files, 3rd Generation, Sub-fonds 3, Box 12-418-412. Hale to Payne, May 5, 1957.

WHO Archives. ARC010-3, Centralized Files, 3rd Generation, Sub-fonds 3, Box 12-418-412. Memo to All Influenza Centres, May 24, 1957.

WHO Archives. ARC010-3, Centralized Files, 3rd Generation, Sub-fonds 3, Box L2-308-302.

WHO Archives. ARC010-3, Centralized Files, 3rd Generation, Sub-fonds 3, Box Z2-181-44. Kaplan to Laver, May 14, 1970.

WHO Archives. ARC010-3, Centralized Files, 3rd Generation, Sub-fonds 3, Box Z2-181-49. Webster to Cockburn, August 20, 1970.

WHO Archives. ARC010-3, Centralized Files, 3rd Generation, Sub-fonds 3, Box Z2-181-118(A).

WHO Archives. ARC010-3, Centralized Files, 3rd Generation, Sub-fonds 3, Box Z2-180-11. "Ecology of Influenza: Framework for a Global Programme," 1982.

WHO Archives. ARC010-3, Centralized Files, 3rd Generation, Sub-fonds 3, Box Z2-180-11. Guo Yuanji to K. Esteves, February 7, 1985.

WHO Archives. ARC010-3, Centralized Files, 3rd Generation, Sub-fonds, Box Z2-181-144.

WHO Archives. ARC010-3, Centralized Files, 3rd Generation, Sub-fonds 3, CTS Agreement with the John Curtin School of Medical Research, Canberra, Australia, in respect of ecological study of influenza, January 12, 1973.

"WHO Blasts China for Withholding Bird Flu Samples." *New Scientist* (November 3, 2006). www.newscientist.com/article/dn10439-who-blasts-china-for-withholding -bird-flu-samples.

Wolfe, Nathan. *The Viral Storm: The Dawn of a New Pandemic Age.* New York: Times Books, 2011.

Zhou, Peng, Xing-Lou Yang, Xian-Guang Wang, Ben Hu, Lei Zhang, Wei Zhang, Hao-Rui Si, et al. "A Pneumonia Outbreak Associated with a New Coronavirus of Probable Bat Origin." *Nature*, February 3, 2020. doi.org/10.1038/s41586-020 -2012-7.

Zhu, Jiming (see also Chi-Ming Chu), Jun Xiao, and Chengzhang Hao [朱既明, 萧俊, 郝成章]。1957 年流行性感冒流行的病毒类型和性状 I. 长春 分离的病毒, 《科学通报》, June 30, 1957.

"Zhu Jiming."《朱既明》—中国科学技术专家传略·医学编 基础医学卷 一—中国 工具书网络出版传略.

SCHOLARLY AND SECONDARY WORKS

Adams, Vincanne. *Metrics: What Counts in Global Health.* Princeton, NJ: Princeton University Press, 2016.

Adams, Vincanne, Michelle Murphy, and Adele E. Clarke. "Anticipation: Technoscience, Life, Affect, Temporality." *Subjectivity* 28, no. 1 (2009): 246–65.

Adams, Vincanne, Thomas E. Novotny, and Hannah Leslie. "Global Health Diplomacy." *Medical Anthropology* 27, no. 4 (2008): 315–23.

Anagnost, Ann. "The Corporeal Politics of Quality (*Suzhi*)." *Public Culture* 16, no. 2 (2004): 189–208.

Anagnost, Ann. *National Past-Times: Narrative, Representation, and Power in Modern China.* Durham, NC: Duke University Press, 1997.

Anderson, Warwick. *Colonial Pathologies: American Tropical Medicine, Race, and Hygiene in the Philippines.* Durham, NC: Duke University Press, 2006.

Anderson, Warwick. "Natural Histories of Infectious Disease: Ecological Vision in Twentieth-Century Biomedical Science." *Osiris* 19 (2004): 39–61.

Anderson, Warwick. "Nowhere to Run, Rabbit: The Cold-War Calculus of Disease Ecology." *History and Philosophy of the Life Sciences* 39, no. 2 (June 2017): 13. doi.org /10.1007/s40656-017-0140-7.

Anderson, Warwick. "Postcolonial Ecologies of Parasite and Host: Making Parasitism Cosmopolitan." *Journal of the History of Biology* 49, no. 2 (2016): 241–59.

Andreas, Joel. *Rise of the Red Engineers: The Cultural Revolution and the Origins of China's New Class.* Stanford, CA: Stanford University Press, 2009.

Arnold, David, ed. *Warm Climates and Western Medicine: The Emergence of Tropical Medicine, 1500–1900.* Amsterdam: Editions Rodopi, 1996.

Auewarakul, Prasert, Wanna Hanchaoworakul, and Kumnuan Ungchusak. "Institutional Responses to Avian Influenza in Thailand: Control of Outbreaks in Poultry and Preparedness in the Case of Human-to-Human Transmission." *Anthropology and Medicine* 15, no. 1 (April 1, 2008): 61–67. doi:10.1080/13648470801919065.

Beck, Ulrich. *Risk Society: Towards a New Modernity*. London: SAGE, 1992.

Beck, Ulrich. *World Risk Society*. Cambridge: Polity, 2009.

Belser, Jessica A., Kristy J. Szretter, Jacqueline M. Katz, and Terrence M. Tumpey. "Use of Animal Models to Understand the Pandemic Potential of Highly Pathogenic Avian Influenza Viruses." *Advances in Virus Research* 73 (2009): 55–97. doi.org/10 .1016/S0065-3527(09)73002-7.

Benson, Etienne. *Wired Wilderness: Technologies of Tracking and the Making of Modern Wildlife*. Baltimore: Johns Hopkins University Press, 2010.

Biehl, João Guilherme, and Adriana Petryna. *When People Come First: Critical Studies in Global Health*. Princeton, NJ: Princeton University Press, 2013.

Boltanski, Luc, Ève Chiapello, and Gregory Elliott. *The New Spirit of Capitalism*. London: Verso, 2018.

Bourdieu, Pierre. *The Logic of Practice*. Stanford, CA: Stanford University Press, 1990.

Boyd, William, and Michael Watts. "Agro-industrial Just-in-Time: The Chicken Industry and Postwar American Capitalism." In *Globalising Food: Agrarian Questions and Global Restructuring*, edited by David Goodman and Michael Watts. Hoboken, NJ: Taylor and Francis, 2013.

Bresalier, Michael. "Fighting Flu: Military Pathology, Vaccines, and the Conflicted Identity of the 1918–19 Pandemic in Britain." *Journal of the History of Medicine and Allied Sciences* 68, no. 1 (January 1, 2013), 124. doi.org/10.1093/jhmas/jrr041.

Bresalier, Michael. "Neutralizing Flu: 'Immunological Devices' and the Making of a Virus Disease." In *Crafting Immunity: Working Histories of Clinical Immunology*, edited by Pauline Mazumdar, Kenton Kroker, and Jennifer Keelan. London: Ashgate, 2008.

Bresalier, Michael. "Sharing Viruses and Vaccines: Economies of Exchange in Global Influenza Control, 1947–1957." Unpublished paper presented at Universite de Strasbourg, December 6–7, 2012.

Bresalier, Michael. "Uses of a Pandemic: Forging the Identities of Influenza and Virus Research in Interwar Britain." *Social History of Medicine* 25, no. 2 (2011): 401. doi .org/10.1093/shm/hkr162.

Briggs, Charles L., and Mark Nichter. "Biocommunicability and the Biopolitics of Pandemic Threats." *Medical Anthropology* 28, no. 3 (2009): 189–98.

Brown, Theodore M., Marcos Cueto, and Elizabeth Fee. "The World Health Organization and the Transition from 'International' to 'Global' Public Health." *American Journal of Public Health* 96, no. 1 (January 2006): 62. doi.org/10.2105 /AJPH.2004.050831.

Bu, Liping. "Anti-malaria Campaigns and the Socialist Reconstruction of China, 1950–1980." *East Asian History* 39 (2014): 117–30.

Caduff, Carlo. "After the Next: Notes on Serial Novelty." *Medicine Anthropology Theory* 5, no. 4 (September 10, 2018): 86–105. doi.org/10.17157/mat.5.4.623.

Caduff, Carlo. *The Pandemic Perhaps: Dramatic Events in a Public Culture of Danger*. Berkeley: University of California Press, 2015.

Callon, Michel, and Vololona Rabeharisoa. "Research 'in the Wild' and the Shaping of New Social Identities." *Technology in Society* 25, no. 2 (2003): 193–204.

Canguilhem, Georges. "The History of Science." In *A Vital Rationalist: Selected Writings from Georges Canguilhem*, edited by Francois Delaporte. New York: Zone, 1994.

Canguilhem, Georges. "The Living and Its Milieu." In *Science, Reason, Modernity: Readings for an Anthropology of the Contemporary*, edited by Anthony Stavrianakis, Gaymon Bennett, and Lyle Fearnley, 168–90. New York: Fordham University Press, 2015.

Chen, Nancy N., and Lesley Alexandra Sharp. *Bioinsecurity and Vulnerability*. Santa Fe, NM: School for Advanced Research Press, 2014.

Cherry, P., and T. R Morris. *Domestic Duck Production: Science and Practice*. Cambridge, MA: CABI, 2008.

Chien, Yu-Ju. "How Did International Agencies Perceive the Avian Influenza Problem? The Adoption and Manufacture of the 'One World, One Health' Framework." *Sociology of Health and Illness* 35, no. 2 (February 2013): 213–26. doi.org/10.1111/j.1467-9566.2012.01534.x.

Coggins, Chris. *The Tiger and the Pangolin: Nature, Culture and Conservation in China*. Honolulu: University of Hawai'i Press, 2004.

Collier, Stephen J. "Global Assemblages." *Theory, Culture and Society* 23 (2006).

Collier, Stephen J. *Post-Soviet Social: Neoliberalism, Social Modernity, Biopolitics*. Princeton, NJ: Princeton University Press, 2011.

Collier, Stephen J., and Andrew Lakoff. "Vital Systems Security: Reflexive Biopolitics and the Government of Emergency." *Theory, Culture and Society* 32, no. 2 (2015): 19–51.

Collier, Stephen J., and Aihwa Ong. "Global Assemblages, Anthropological Problems." In *Global Assemblages: Technology, Politics, and Ethics as Anthropological Problems*, edited by Aihwa Ong and Stephen J. Collier. Malden, MA: Blackwell, 2005.

Cooper, Melinda. "Pre-empting Emergence: The Biological Turn in the War on Terror." *Theory, Culture and Society* 23, no. 4 (July 2006): 113–35. doi.org/10.1177/0263276406065121.

Craddock, Susan, and Steve Hinchliffe. "One World, One Health? Social Science Engagements with the One Health Agenda." *Social Science and Medicine* 129 (2015): 2. doi.org/10.1016/j.socscimed.2014.11.016.

Crosby, Alfred W. *America's Forgotten Pandemic: The Influenza of 1918*. Cambridge: Cambridge University Press, 2003.

Cunningham, Andrew. "Transforming Plague: The Laboratory and the Identity of Infectious Disease." In *The Laboratory Revolution in Medicine*, edited by Andrew Cunningham and Perry Williams. Cambridge: Cambridge University Press, 2002.

Davis, Mike. *The Monster at Our Door: The Global Threat of Avian Flu*. New York: New Press, 2005.

Dehner, George. *Influenza: A Century of Science and Public Health Response*. Pittsburgh: University of Pittsburgh Press, 2012.

Delanda, Manuel. *A New Philosophy of Society: Assemblage Theory and Social Complexity*. London: Continuum, 2006.

Delaporte, François. *Chagas Disease: History of a Continent's Scourge*. New York: Fordham University Press, 2012.

Deleuze, Gilles, and Félix Guattari. "10,000 BC: The Geology of Morals (Who Does the Earth Think It Is?)." In *A Thousand Plateaus*, translated by Brian Massumi. Minneapolis: University of Minnesota Press, 1987.

Deleuze, Gilles, and Félix Guattari. *A Thousand Plateaus*, translated by Brian Massumi. Minneapolis: University of Minnesota Press, 1987.

Douglas, Mary. *Purity and Danger: An Analysis of Concepts of Pollution and Taboo*. New York: Routledge, 2015.

Downey, Gary Lee, and Joseph Dumit. *Cyborgs and Citadels: Anthropological Interventions in Emerging Sciences and Technologies*. Santa Fe, NM: School of American Research Press, 1997.

Dunk, James H., David S. Jones, Anthony Capon, and Warwick H. Anderson. "Human Health on an Ailing Planet—Historical Perspectives on Our Future." Edited by Debra Malina. *New England Journal of Medicine* 381, no. 8 (August 22, 2019): 778–82. doi.org/10.1056/NEJMms1907455.

Dunn, Elizabeth C. "Trojan Pig: Paradoxes of Food Safety Regulation." *Environment and Planning A*, 35 (2003): 1493–511.

Edwards, Paul N. *A Vast Machine: Computer Models, Climate Data, and the Politics of Global Warming*. Cambridge, MA: MIT Press, 2010.

Ellis, Linden J., and Jennifer L. Turner. "Where the Wild Things Are . . . Sold." *China Environment Series* 9 (2007): 131–34. www.wilsoncenter.org/publication/full-publication-1.

Eyler, John M. "De Kruif's Boast: Vaccine Trials and the Construction of a Virus." *Bulletin of the History of Medicine* 80, no. 3 (2006): 409–38. doi.org/10.1353/bhm.2006.0092.

Farquhar, Judith. *Appetites: Food and Sex in Postsocialist China*. Durham, NC: Duke University Press, 2002.

Farquhar, Judith. "Market Magic: Getting Rich and Getting Personal in Medicine after Mao." *American Ethnologist* 23, no. 2 (1996): 239–57.

Fassin, Didier, and Mariella Pandolfi. "Introduction: Military and Humanitarian Government in the Age of Intervention." In *Contemporary States of Emergency: The Politics of Military and Humanitarian Inventions*, edited by Didier Fassin and Mariella Pandolfi. Cambridge, MA: Zone, 2010.

Fausto, C., and L. Costa. "Feeding (and Eating): Reflections on Strathern's 'Eating (and Feeding).'" *Cambridge Anthropology* 31, no. 1 (2013): 156–62.

Fearnley, Lyle. "Epidemic Intelligence: Langmuir and the Birth of Disease Surveillance." *Behemoth* 3, no. 3 (2010): 37–56.

Fearnley, Lyle. "Signals Come and Go: Syndromic Surveillance and Styles of Biosecurity." *Environment and Planning A* 40, no. 7 (2008): 1615–32.

Fidler, David P. "Germs, Governance, and Global Public Health in the Wake of SARS." *Journal of Clinical Investigation* 113, no. 6 (March 15, 2004): 799. doi.org/10.1172/JCI200421328.

Fidler, David P. "Influenza Virus Samples, International Law, and Global Health Diplomacy." *Emerging Infectious Diseases* 14, no. 1 (2008): 88–94.

Fischer, M. M. J., "A Tale of Two Genome Institutes: Qualitative Networks, Charismatic Voice, and R&D Strategies—Juxtaposing GIS Biopolis and BGI." *Science, Technology and Society* 23, no. 2 (2018): 271–88.

Fischer, Michael M. J. *Anthropological Futures*. Durham, NC: Duke University Press, 2009.

Fischer, Michael M. J. "Biopolis: Asian Science in the Global Circuitry." *Science, Technology and Society* 18, no. 3 (2013): 381–406.

Fisher, P. "The Pixel: A Snare and a Delusion." *International Journal of Remote Sensing* 18, no. 3 (February 1, 1997): 679–85. doi.org/10.1080/014311697219015.

Fleck, Ludwik. *Genesis and Development of a Scientific Fact*, edited by Thaddeus J. Trann and Robert T. Merton. Chicago: University of Chicago Press, 1981.

Fleming, James Rodger. "Planetary-Scale Fieldwork: Harry Wexler on the Possibilities of Ozone Depletion and Climate Control." In *Knowing Global Environments: New Historical Perspectives on the Field Sciences*, ed. Jeremy Vetter. New Brunswick, NJ: Rutgers University Press, 2010.

Forster, Paul, and Olivier Charnoz. "Producing Knowledge in Times of Health Crises: Insights from the International Response to Avian Influenza in Indonesia." *Revue d'anthropologie des conaissances* 7, no. 1 (2013): 112–44.

Fortun, Kim. "Ethnography in/of/as Open Systems." *Reviews in Anthropology* 32, no. 2 (January 2003): 171–90.

Fortun, Kim. "Ethnography in Late Industrialism." *Cultural Anthropology* 27, no. 3 (2012): 446–64.

Franklin, Sarah, Celia Lury, and Jackie Stacey. *Global Nature, Global Culture*. London: SAGE, 2000.

Fujimura, Joan. "Crafting Science: Standardized Packages, Boundary Objects, and 'Translation.'" In *Science as Practice and Culture*, edited by Andrew Pickering, 168–211. Chicago: University of Chicago Press, 1992.

Gell, Alfred. "The Technology of Enchantment and the Enchantment of Technology." In *The Art of Anthropology: Essays and Diagrams*, edited by Eric Hirsch. London: The Athlone Press, 1999.

Goodman, David, Bernardo Sorj, and John Wilkinson. *From Farming to Biotechnology: A Theory of Agro-industrial Development*. New York: Basil Blackwell, 1987.

Greenfeld, Karl Taro. "Wild Flavor." *Paris Review* 47, no. 175 (2005): 7.

Gross, Miriam. *Farewell to the God of Plague: Chairman Mao's Campaign to Deworm China*. Oakland: University of California Press, 2016.

Hacking, Ian. *The Taming of Chance*. Cambridge: Cambridge University Press, 1990.

Haraway, Donna. *Staying with the Trouble: Making Kin in the Cthulucene*. Durham, NC: Duke University Press, 2016.

Harris, Phil. "Avian Influenza: An Animal Health Issue." August 2006. www.fao.org /avianflu/en/issue.html.

Harrison, Mark. "Pandemics." In *The Routledge History of Disease*, edited by Mark Jackson, 129–46. London: Routledge, 2017.

Hathaway, Michael J. *Environmental Winds: Making the Global in Southwest China*. Berkeley: University of California Press, 2013.

Hayden, Cori. *When Nature Goes Public: The Making and Unmaking of Bioprospecting in Mexico*. Princeton, NJ: Princeton University Press, 2003.

Helmreich, Stefan. *Alien Ocean: Anthropological Voyages in Microbial Seas*. Berkeley: University of California Press, 2009.

Hinchliffe S., S. Lavau, J. Allen, N. Bingham, and S. Carter. "Biosecurity and the Topologies of Infected Life: From Borderlines to Borderlands." *Transactions of the Institute of British Geographers* 38, no. 4 (2013): 531–43.

Hinchliffe, Stephen, and Nick Bingham. "Securing Life: The Emerging Practices of Biosecurity." *Environment and Planning A* 40, no. 7 (2008): 1534–51.

Hinterberger, Amy, and Natalie Porter. "Genomic and Viral Sovereignty: Tethering the Materials of Global Biomedicine." *Public Culture* 27, no. 2–76 (2015): 361–86. doi.org/10.1215/08992363-2841904.

Hirzel, Alexandre H., and Gwenaëlle Le Lay. "Habitat Suitability Modelling and Niche Theory." *Journal of Applied Ecology* 45, no. 5 (October 1, 2008): 1372–81. doi:10.1111/j.1365-2664.2008.01524.x.

Honigsbaum, Mark. *A History of the Great Influenza Pandemics: Death, Panic and Hysteria, 1830–1920*. London: I. B. Tauris, 2014.

Hu, Zhanping, Qian Forrest Zhang, and John A. Donaldson. "Farmers' Cooperatives in China: A Typology of Fraud and Failure." *The China Journal* 78 (2017): 1–24.

Huang, Hsing-Tsung, and Joseph Needham. *Science and Civilisation in China*. Vol. 6, *Biology and Biological Technology*. Cambridge: Cambridge University Press, 2000.

Huang, Hsuan-Ying. "Untamed Jianghu or Emerging Profession: Diagnosing the Psycho-boom amid China's Mental Health Legislation." *Culture, Medicine, and Psychiatry* 42, no. 2 (June 2018): 371–400. doi.org/10.1007/s11013-017-9553-8.

Huang, Jikun, Xiaobing Wang, and Guangguang Qiu. *Small-Scale Farmers in China in the Face of Modernisation and Globalisation*. London: IIED/HIVOS, 2012, 16.

Huang, Phillip C. C. "China's New-Age Small Farms and Their Vertical Integration: Agribusiness or Co-ops?" *Modern China* 32, no. 2 (2011): 107–34.

Huang, Yu. "Neoliberalizing Food Safety Control: Training Licensed Fish Veterinarians to Combat Aquaculture Drug Residues in Guangdong." *Modern China* 42, no. 5 (September 2016): 535–65. doi.org/10.1177/0097700415605322.

Hvistendahl, Mara. "Veterinarian-in-Chief." *Science* 341, no. 6142 (2013): 122–25.

Ingold, Tim. *Being Alive: Essays on Movement, Knowledge and Description*. London: Routledge, 2011.

Ingold, Tim. *Hunters, Pastoralists and Ranchers: Reindeer Economies and Their Transformations*. Cambridge: Cambridge University Press, 1980.

Ingold, Tim. "Making Things, Growing Plants, Raising Animals, and Bringing Up Children." In *Perception of the Environment: Essays on Livelihood, Dwelling, and Skill*, by Tim Ingold. London: Routledge, 2000.

Jullien, Francois. *Detour and Access: Strategies of Meaning in China and Greece*. New York: Zone, 2000.

Kaufman, Joan. "China's Heath Care System and Avian Influenza Preparedness." *Journal of Infectious Diseases* 197, Suppl. 1 (2008): s8.

Kaufman, Joan. "SARS and China's Health-Care Response: Better to Be Both Red and Expert!" In *SARS in China: Prelude to Pandemic?*, edited by Arthur Kleinman and James L. Watson. Stanford, CA: Stanford University Press, 2006.

Ke, Bingsheng, and Yijun Han. "Poultry Sector in China: Structural Change during the Past Decade and Future Trends." Food and Agriculture Organization of the United Nations, 2008.

Keck, Frédéric. "From Purgatory to Sentinel: 'Forms/Events' in the Field of Zoonoses." *Cambridge Anthropology* 32, no. 1 (2014): 47–61.

Kerner, Susanne, Cynthia Chou, and Morten Warmind. *Commensality: From Everyday Food to Feast*. New York: Bloomsbury, 2015.

King, Nicholas B. "Security, Disease, Commerce: Ideologies of Postcolonial Global Health." *Social Studies of Science* 32, nos. 5–6 (2002): 763–89.

Kipnis, Andrew B. *Producing Guanxi: Sentiment, Self, and Subculture in a North China Village*. Durham, NC: Duke University Press, 1997.

Kipnis, Andrew B. "*Suzhi*: A Keyword Approach." *China Quarterly* 186, no. 1 (2006): 295–313.

Klein, Jakob. "Connecting with the Countryside? 'Alternative' Food Movements with Chinese Characteristics." In *Ethical Eating in the Postsocialist and Socialist World*, edited by Yuson Jung, Jakob A. Klein, and Melissa L. Caldwell. Berkeley: University of California Press, 2014.

Kleinman, A. M., B. R. Bloom, A. Saich, K. A. Mason, and F. Aulino. "Asian Flus in Ethnographic and Political Context: A Biosocial Approach." *Anthropology and Medicine* 15, no. 1 (2008): 1–5. dx.doi.org/10.1080/13648470801918968.

Kleinman, Arthur, and James L. Watson. "Introduction: SARS in Social and Historical Context." In *SARS in China: Prelude to Pandemic?*, edited by Arthur Kleinman and James L. Watson. Stanford, CA: Stanford University Press, 2006.

Knorr-Cetina, Karin. "The Couch, the Cathedral, and the Laboratory: On the Relationship between Experiment and Laboratory in Science." In *Science as Practice and Culture*, edited by Andrew Pickering. Chicago: University of Chicago Press, 1992.

Knorr-Cetina, Karin. *Epistemic Cultures: How the Sciences Make Knowledge*. Cambridge, MA: Harvard University Press, 2009.

Knorr-Cetina, Karin. "Laboratory Studies: The Cultural Approach to the Study of Science." In *Handbook of Science and Technology Studies*, edited by Sheila Jasanoff, Gerald E. Markle, James C. Peterson, and Trevor Pinch, 140–66. Thousand Oaks, CA: SAGE, 1995.

Knorr-Cetina, Karin. *The Manufacture of Knowledge: An Essay on the Constructivist and Contextual Nature of Science*. Oxford: Pergamon, 1981.

Kohler, Robert. *Landscapes and Labscapes: Exploring the Lab-Field Border in Biology*. Chicago: University of Chicago Press, 2002.

Kohler, Robert. "Prospects." In *Knowing Global Environments: New Historical Perspectives on the Field Sciences*, edited by Jeremy Vetter. New Brunswick, NJ: Rutgers University Press, 2010.

Kuhn, Thomas S. *The Structure of Scientific Revolutions*. Chicago: University of Chicago Press, 1996.

Kuklick, Henrika, and Robert E. Kohler. "Science in the Field." *Osiris* 11 (1996): 1–265.

Kuo, Wen-Hua. "The Voice on the Bridge: Taiwan's Regulatory Engagement with Global Pharmaceuticals." *East Asian Science, Technology and Society* 3, no. 1 (March 1, 2009): 51–72. doi.org/10.1215/s12280-008-9066-1.

Kwa, Chunglin. "Local Ecologies and Global Science: Discourses and Strategies of the International Geosphere-Biosphere Programme." *Social Studies of Science* 35, no. 6 (2005): 923–50.

Lai, Lili. "Everyday Hygiene in Rural Henan." *Positions* 22, no. 3 (2014): 635–59.

Lakoff, Andrew. "The Generic Biothreat, or, How We Became Unprepared." *Cultural Anthropology* 23, no. 3 (2008): 399–428.

Lakoff, Andrew. "Two Regimes of Global Health." *Humanity* 1, no. 1 (2010): 59–79.

Lakoff, Andrew. *Unprepared: Global Health in a Time of Emergency*. Oakland: University of California Press, 2017.

Lakoff, Andrew, and Stephen J. Collier. *Biosecurity Interventions: Global Health and Security in Question*. New York: Columbia University Press, 2008.

Lamoreaux, Janelle. "What if the Environment Is a Person? Lineages of Epigenetic Science in a Toxic China." *Cultural Anthropology* 31, no. 2 (May 4, 2016): 188–214. doi.org/10.14506/ca31.2.03.

Landecker, Hannah. "Antibiotic Resistance and the Biology of History." *Body and Society* 22, no. 4 (2016): 19–52.

Landecker, Hannah. "Food as Exposure: Nutritional Epigenetics and the New Metabolism." *BioSocieties* 6, no. 2 (2011): 167–94.

Landecker, Hannah. "The Matter of Practice in the Historiography of the Experimental Life Sciences." In *Handbook of the Historiography of Biology*, vol. 1, edited by Michael Dietrich, Mark Borrello, and Oren Harman. Cham, Switzerland: Springer International, 2018.

Lansing, Stephen J. *Priests and Programmers: Technologies of Power in the Engineered Landscape of Bali*. Princeton, NJ: Princeton University Press, 2009.

Latour, Bruno. *Facing Gaia: Eight Lectures on the New Climatic Regime*, translated by Catherine Porter. Cambridge: Polity, 2018.

Latour, Bruno. "Give Me a Laboratory and I Will Raise the World." In *Science Observed: Perspectives on the Social Study of Science*, edited by Karin D. Knorr-Cetina and Michael Mulkay, 141–69. London: SAGE, 1983.

Latour, Bruno. "On Technical Mediation—Philosophy, Sociology, Geneaology." *Common Knowledge* 3, no. 2 (1994): 29–64.

Latour, Bruno. *Pandora's Hope: Essays on the Reality of Science Studies*. Cambridge, MA: Harvard University Press, 1999.

Latour, Bruno. *The Pasteurization of France*. Cambridge, MA: Harvard University Press, 1988.

Latour, Bruno. *Science in Action: How to Follow Scientists and Engineers through Society*. Cambridge, MA: Harvard University Press, 1987.

Latour, Bruno, and Steve Woolgar. *Laboratory Life: The Construction of Scientific Facts*. Princeton, NJ: Princeton University Press, 1986.

Law, John. "Disaster in Agriculture: Or Foot and Mouth Mobilities." *Environment and Planning A* 38, no. 2 (February 1, 2006): 227–39. doi.org/10.1068/a37273.

Law, John. "On the Methods of Long Distance Control: Vessels, Navigation, and the Portuguese Route to India." In *Power, Action, and Belief: A New Sociology of Knowledge?*, edited by John Law. London: Routledge, 1987.

Leach, Edmund. "Anthropological Aspects of Language: Animal Categories and Verbal Abuse." *Anthrozoös* 2, no. 3 (1989): 151–65.

Leach, Edmund. *Political Systems of Highland Burma: A Study of Kachin Social Structure.* Boston: Beacon, 1965.

Lee, Kelley. *The World Health Organization (WHO).* London: Routledge, 2009.

Legendre, Pierre, and Marie Josée Fortin. "Spatial Pattern and Ecological Analysis." *Vegetatio* 80, no. 2 (June 1989): 107–38.

Lei, Xianglin. *Neither Donkey nor Horse: Medicine in the Struggle over China's Modernity.* Chicago: University of Chicago Press, 2014.

Le Roy Ladurie, Emmanual. "A Concept: The Unification of the Globe by Disease." In *The Mind and Method of the Historian*, by Emmanual Le Roy Ladurie. Chicago: University of Chicago Press, 1981.

Li, Chengyue, Yilin Hou, Mei Sun, Jun Lu, Ying Wang, Xiaohong Li, Fengshui Chang, and Mo Hao. "An Evaluation of China's New Rural Cooperative Medical System: Achievements and Inadequacies from Policy Goals." *BMC Public Health* 15 (2015): 1079. doi.org/10.1186/s12889-015-2410-1.

Li, Hui. "The Countryside Vets the State Left Behind." *Sixth Tone*, June 14, 2018. www.sixthtone.com/news/1002439/The%20Countryside%20Vets%20the%20State%20Left%20Behind.

Li, Jiao. "In China's Backcountry, Tracking Lethal Bird Flu." *Science* 330, no. 6002 (2010): 313.

Liang, Ruiha [梁瑞华]. 高致病性禽流感疫病控制模式与宏观管理研究. 武汉: 武汉理工大学博士学位论文, 2007.

Lin, Wen-Yuan. "Displacement Agency: The Enactment of Patients' Agency in and beyond Haemodialysis Practices." *Science, Technology and Human Values* 38, no. 3 (2012): 421–43.

Lingohr, Susanne. "Rural Households, Dragon Heads and Associations: A Case Study of Sweet Potato Processing in Sichuan Province." *China Quarterly* 192 (2007): 898–914.

Liu, Tik-sang. "Custom, Taste and Science: Raising Chickens in the Pearl River Delta Region, South China." *Anthropology and Medicine* 15, no. 1 (April 1, 2008): 7–18. doi:10.1080/13648470801918992.

Livingstone, David N. *Putting Science in Its Place: Geographies of Scientific Knowledge.* Chicago: University of Chicago Press, 2013.

Lora-Wainwright, Anna. *Fighting for Breath: Living Morally and Dying of Cancer in a Chinese Village.* Honolulu: University of Hawai'i Press, 2013.

Lora-Wainwright, Anna. "Of Farming Chemicals and Cancer Deaths: The Politics of Health in Contemporary Rural China." *Social Anthropology* 17, no. 1 (2009): 56–73.

Lowe, Celia. "Viral Clouds: Becoming H5N1 in Indonesia." *Cultural Anthropology* 25, no. 4 (2010): 625–49.

Lowe, Celia. *Wild Profusion: Biodiversity Conservation in an Indonesian Archipelago.* Princeton, NJ: Princeton University Press, 2006.

Lynteris, Christos. *Ethnographic Plague: Configuring Disease on the Chinese-Russian Frontier.* London: Palgrave Macmillan, 2016.

Lynteris, Christos. "Zoonotic Diagrams: Mastering and Unsettling Human–Animal Relations." *Journal of the Royal Anthropological Institute* 23, no. 3 (2017): 463–85.

MacPhail, Theresa. "The Politics of Bird Flu: The Battle over Viral Samples and China's Role in Global Public Health." *Journal of Language and Politics* 8, no. 3 (2009): 456–75.

MacPhail, Theresa. *The Viral Network: A Pathography of the H1N1 Influenza Pandemic.* Ithaca, NY: Cornell University Press, 2014.

Malinowski, Bronislaw. *Coral Gardens and Their Magic.* Vol. 1, *Soil-Tilling and Agricultural Rites in the Trobriand Islands.* Bloomington: Indiana University Press, 1965.

Marcus, George E. "Ethnography in/of the World System: The Emergence of Multisited Ethnography." *Annual Review of Anthropology* 24 (1995): 95–117.

Marx, Karl. *Capital,* vol. 2. London: Penguin, 1978.

Mason, Katherine. *Infectious Change: Reinventing Chinese Public Health after an Epidemic.* Palo Alto, CA: Stanford University Press, 2016.

Mason, Katherine A. "To Your Health! Toasting, Intoxication and Gendered Critique among Banqueting Women." *China Journal* 69 (2013): 108–33.

McKenna, Maryn. "The Race to Find the Next Pandemic—Before It Finds Us." *Wired,* April 12, 2018. www.wired.com/story/the-race-to-find-the-next-pandemic-before -it-finds-us.

Mei Fuchun. *Research on the Government's Use of Culling Compensation Policy in Response to the Avian Influenza Crisis* 《政府应对禽流感突发事件的扑杀补偿政策研究》 中国农业出版社, 2011.

Mendelsohn, Andrew J. "From Eradication to Equilibrium: How Epidemics Became Complex after World War I." In *Greater Than the Parts: Holism in Biomedicine, 1920–1950,* edited by C. Lawrence and G. Weisz, 303–31. Oxford: Oxford University Press, 1998.

Müller-Wille, Staffan, and Hans-Jörg Rheinberger. *A Cultural History of Heredity.* Chicago: University of Chicago Press, 2012.

Newman, Scott H., Boripat Siriaroonat, and Xiangming Xiao. "A One Health Approach to Understanding Dynamics of Avian Influenza in Poyang Lake, China." Kunming, China, Ecohealth, 2012.

Normile, Dennis. "Is China Coming Clean on Bird Flu?" *Science* 314, no. 5801 (2006): 905.

Nowotny, Helga, Peter Scott, and Michael Gibbons. *Re-thinking Science: Knowledge and the Public in an Age of Uncertainty.* Cambridge: Polity, 2001.

Ong, Aihwa. *Flexible Citizenship: The Cultural Logics of Transnationality.* Durham, NC: Duke University Press, 1999.

Ong, Aihwa. *Fungible Life: Experiment in the Asian City of Life.* Durham, NC: Duke University Press, 2016.

Ong, Aihwa. "Introduction: An Analysis of Biotechnology and Ethics at Multiple Scales." In *Asian Biotech: Ethics and Communities of Fate,* edited by Aihwa Ong and Nancy N. Chen. Durham, NC: Duke University Press, 2010.

Osburg, John. *Anxious Wealth: Money and Morality among China's New Rich*. Stanford, CA: Stanford University Press, 2013.

Ouagrham-Gormley, Sonia. "Growth of the Anti-plague System during the Soviet Period." *Critical Reviews in Microbiology* 32, no. 1 (2006): 33–46.

Oxfeld, Ellen. *Drink Water, but Remember the Source: Moral Discourse in a Chinese Village*. Berkeley: University of California Press, 2010.

Packard, Randall M. *A History of Global Health: Interventions into the Lives of Other Peoples*. Baltimore: Johns Hopkins University Press, 2016.

Padmawati, Siwi, and Mark Nichter. "Community Response to Avian Flu in Central Java, Indonesia." *Anthropology and Medicine* 15, no. 1 (April 1, 2008): 31–51. doi.10.1080/13648470801919032.

Paxson, Heather. *The Life of Cheese: Crafting Food and Value in America*. Berkeley: University of California Press, 2012.

Peckham, Robert, and Ria Sinha. "Satellites and the New War on Infection: Tracking Ebola in West Africa." *Geoforum* 80, Suppl. C (March 1, 2017): 24–38.

Pickering, Andrew. *The Mangle of Practice: Time, Agency and Science*. Chicago: University of Chicago Press, 1995.

Porter, Dorothy. "Stratification and Its Discontents: Professionalization and Conflict in the British Public Health Service, 1848–1914." In *A History of Education in Public Health: Health That Mocks the Doctors' Rules*, edited by Elizabeth Fee and Roy Acheson. Oxford: Oxford University Press, 1991.

Porter, Natalie. "Bird Flu Biopower: Strategies for Multispecies Coexistence in Việt Nam." *American Ethnologist* 40, no. 1 (2013): 132–48.

Porter, Theodore M. *The Rise of Statistical Thinking: 1820-1900*. Princeton, NJ: Princeton University Press, 2011.

Preston, Richard. *The Hot Zone: The Terrifying True Story of the Origins of the Ebola Virus*. New York: Anchor, 1995.

Qu, Yanchun. "Jitihua shiqi de nongcun gonggongpin gongji: jingyan yu jiejian" [Rural public goods supply of the people's commune period: experiences and lessons]. *Anhui 'nongye kexue* 39, no. 3 (2011): 1782–84.

Quammen, David. *Spillover: Animal Infections and the Next Human Pandemic*. New York: W. W. Norton, 2012.

Rabinow, Paul. "Artificiality and Enlightenment." In *Essays on the Anthropology of Reason*, by Paul Rabinow, 91–111. Princeton, NJ: Princeton University Press, 1996.

Rabinow, Paul. *French DNA: Trouble in Purgatory*. Chicago: University of Chicago Press, 1999.

Rabinow, Paul. *Making PCR: A Story of Biotechnology*. Chicago: University of Chicago Press, 1996.

Rabinow, Paul, and Gaymon Bennett. *Designing Human Practices: An Experiment with Synthetic Biology*. Chicago: University of Chicago Press, 2012.

Rabinow, Paul, and William M. Sullivan. *Interpretive Social Science: A Second Look*. Berkeley: University of California Press, 1987.

Redfield, Peter. *Life in Crisis: The Ethical Journey of Doctors without Borders*. Berkeley: University of California Press, 2014.

Rees, Tobias. "Humanity/Plan; or, On the 'Stateless' Today (Also Being an Anthropology of Global Health)." *Cultural Anthropology* 29, no. 3 (August 11, 2014): 457–78.
doi.org/10.14506/ca29.3.02.

Rheinberger, Hans-Jörg. *An Epistemology of the Concrete: Twentieth-Century Histories of
Life*. Durham, NC: Duke University Press, 2010.

Rheinberger, Hans-Jörg. *Toward a History of Epistemic Things: Synthesizing Proteins in the
Test Tube*. Stanford, CA: Stanford University Press, 1997.

Riles, Annelise. *The Network Inside Out*. Ann Arbor: University of Michigan Press,
2000.

Rogaski, Ruth. *Hygienic Modernity: Meanings of Health and Disease in Treaty-Port China*.
Berkeley: University of California Press, 2004.

Safman, Rachel M. "Avian Influenza Control in Thailand: Balancing the Interests of
Different Poultry Producers." In *Avian Influenza: Science, Policy and Politics*, edited
by Ian Scoones. New York: Routledge, 2010.

Samimian-Darash, Limor. "Governing Future Potential Biothreats: Toward an Anthropology of Uncertainty." *Current Anthropology* 54, no. 1 (2013): 1–22.

Santos, Gonçalo. "Rethinking the Green Revolution in South China: Technological
Materialities and Human–Environment Relations." *East Asian Science, Technology
and Society* 5, no. 4 (2011): 495.

Schmalzer, Sigrid. *Green Revolution, Red Revolution: Scientific Farming in Socialist China*.
Chicago: University of Chicago Press, 2016.

Schmitz, Rob. "The Chinese Lake That's Ground Zero for the Bird Flu." *Marketplace:
National Public Radio*. March 30, 2016. www.marketplace.org/2016/03/03/world
/chinese-lake-has-become-ground-zero-bird-flu.

Schneider, Laurence A. *Biology and Revolution in Twentieth-Century China*. Lanham, MD:
Rowman and Littlefield, 2005.

Schneider, Mindi. "Dragon Head Enterprises and the State of Agribusiness in China."
Journal of Agrarian Change 17, no. 1 (2017): 3–21.

Scoones, Ian, and Paul Forster. "The International Response to Highly Pathogenic
Avian Influenza: Science, Policy and Practice." STEPS Working Paper 10, STEPS
Centre, Brighton, UK, 2008.

Scott, James C. *The Moral Economy of the Peasant: Rebellion and Subsistence in Southeast
Asia*. New Haven, CT: Yale University Press, 1976.

Shapin, Steven. "Who Is the Industrial Scientist? Commentary from Academic
Sociology and from the Shop-Floor in the United States, ca. 1900–ca. 1970." In
Knowledge and Social Order: Rethinking the Sociology of Barry Barnes, edited by Mas-
simo Mazzotti. Aldershot, UK: Ashgate, 2008.

Shapin, Steven, and Simon Schaffer. *Leviathan and the Air-Pump: Hobbes, Boyle, and the
Experimental Life*. Princeton, NJ: Princeton University Press, 1985.

Simon, Denis Fred, and Merle Goldman. *Science and Technology in Post-Mao China*. Cambridge, MA: Council on East Asian Studies, 1989.

Sloterdijk, Peter. *In the World Interior of Capital: Towards a Philosophical Theory of Globalization*. Translated by Wieland Hoban. Cambridge: Polity, 2015.

Sloterdijk, Peter. *Spheres*. Vol. 2, *Globes: Macrospherology*. New York: Semiotext(e), 2011.

Staples, Amy L. S. *The Birth of Development: How the World Bank, Food and Agriculture Organization, and World Health Organization Changed the World, 1945-1965*. Kent, OH: Kent State University Press, 2006.

Stavrianakis, Anthony. "From Anthropologist to Actant (and Back to Anthropology): Position, Impasse, and Observation in Sociotechnical Collaboration." *Cultural Anthropology* 30, no. 1 (February 17, 2015): 169–89. doi.org/10.14506/ca30.1.09.

Stephenson, Niamh. "Emerging Infectious Disease/Emerging Forms of Biological Sovereignty." *Science, Technology, and Human Values* 36, no. 5 (2010). doi:10.1177/0162243910388023.

Stevens, Hallam. "Globalizing Genomics: The Origins of the International Nucleotide Sequence Database Collaboration." *Journal of the History of Biology* 51, no. 4 (2018): 657–91.

Stevens, Hallam. *Life out of Sequence: A Data-Driven History of Bioinformatics*. Chicago: University of Chicago Press, 2013.

Strathern, Marilyn. "Eating (and Feeding)." *Cambridge Anthropology* 30, no. 2 (2012): 1–14.

Strathern, Marilyn. "Re-describing Society." *Minerva* 41, no. 3 (2003): 263–76. dx.doi.org /10.1023/A:1025586327342.

Swayne, D. E., and D. L. Suarez. "Highly Pathogenic Avian Influenza." *Review of Science and Technology of the Office Internationale de Epizootiques* 19, no. 2 (2000): 463–82.

Takekawa, J. Y., S. B. Muzaffar, N. J. Hill, D. J. Prosser, S. H. Newman, B. Yan, X. Xiao, et al. "Victims and Vectors: Highly Pathogenic Avian Influenza H5N1 and the Ecology of Wild Birds." *Avian Biology Research* 3, no. 2 (2010): 51–73.

Tambiah, Stanley J. "Animals Are Good to Think and Good to Prohibit." *Ethnology* 8, no. 4 (1969): 423–59.

Thompson, Charis. "When Elephants Stand for Competing Philosophies of Nature: Amboseli National Park, Kenya." In *Complexities: Social Studies of Knowledge Practice*, edited by John Law and Annemarie Mol. Durham, NC: Duke University Press, 2006.

Traweek, Sharon. *Beamtimes and Lifetimes: The World of High Energy Physicists*. Cambridge, MA: Harvard University Press, 1992.

Tsing, Anna Lowenhaupt. *The Mushroom at the End of the World: On the Possibility of Life in Capitalist Ruins*. Princeton, NJ: Princeton University Press, 2015.

Unger, Jonathan. *The Transformation of Rural China*. London: Routledge, 2016.

Vagneron, Frédéric. "Surveiller et s'unir? The Role of WHO in the First International Mobilizations around an Animal Reservoir of Influenza." *Revue d'anthropologie des connaissances* 9, no. 2 (2015): 139–62.

Vetter, Jeremy. "Introduction." In *Knowing Global Environments: New Historical Perspectives on the Field Sciences*, edited by Jeremy Vetter. New Brunswick, NJ: Rutgers University Press, 2010.

Wald, Priscilla. *Contagious: Cultures, Carriers, and the Outbreak Narrative*. Durham, NC: Duke University Press, 2008.

Wallace, Robert. *Big Farms Make Big Flu: Dispatches on Infectious Disease, Agribusiness, and the Nature of Science*. New York: Monthly Review Press, 2016.

Wallace, Robert G., and Rodrick Wallace. *Neoliberal Ebola: Modeling Disease Emergence from Finance to Forest and Farm*. New York: Springer, 2016.

Waller, Emily, Mark Davis, and Niamh Stephenson. "Australia's Pandemic Influenza 'Protect' Phase: Emerging out of the Fog of Pandemic." *Critical Public Health* 26, no. 1 (January 1, 2016): 99–113. doi.org/10.1080/09581596.2014.926310.

Wang, Zuoyue. "Science and the State in Modern China." *Isis* 98 (2007): 558–70.

Weiss, Brad. *Real Pigs: Shifting Values in the Field of Local Pork*. Durham, NC: Duke University Press, 2016.

Whitworth, Darell, Scott Newman, Taej Mundkur, and Phil Harris. "Wild Birds and Avian Influenza: An Introduction to Applied Research and Field Sampling Techniques." FAO Animal Production and Health Manual. Rome, 2007.

Wilson, Marisa, Alfy Gathorne-Hardy, Peter Alexander, and Lisa Boden. "Why 'Culture' Matters for Planetary Health." *Lancet Planetary Health* 2, no. 11 (November 2018): e467–68. doi.org/10.1016/S25425196(18)30205-5.

Wolf, Meike. "Is There Really Such a Thing as 'One Health'? Thinking about a More Than Human World from the Perspective of Cultural Anthropology." *Social Science and Medicine* 129 (March 2015): 5–11. doi.org/10.1016/j.socscimed.2014.06.018.

Wong, Winnie Won Yin. *Van Gogh on Demand: China and the Readymade*. Chicago: University of Chicago Press, 2014.

Woods, Abigail. *A Manufactured Plague: The History of Foot-and-Mouth Disease in Britain*. London: EARTHSCAN, 2004.

Woods, Abigail, Michael Bresalier, Angela Cassidy, and Rachel Mason Dentinger. *Animals and the Shaping of Modern Medicine: One Health and Its Histories*. London: Palgrave Macmillan, 2017.

Worboys, Michael. *Spreading Germs: Disease Theories and Medical Practice in Britain, 1865-1900*. Cambridge: Cambridge University Press, 2000.

Wu, Ziping. "Antibiotic Use and Antibiotic Resistance in Food-Producing Animals in China." OECD Food, Agriculture and Fisheries Papers no. 134. Paris: OECD Publishing, 2019. dx.doi.org/10.1787/4ADba8c1-en.

"Wuhan Virus: Rats and Live Wolf Pups on the Menu at China Food Market Linked to Virus Outbreak." *The Straits Times*, January 22, 2020.

Wynne, Brian. "May the Sheep Safely Graze? A Reflexive View of the Expert–Lay Divide." In *Risk, Environment, and Modernity: Towards a New Ecology*, edited by Scott Lash, Bronislaw Szerszynski, and Brian Wynne. London: SAGE, 1996.

Xie, Chaoping, and Mary A. Marchant. "Supplying China's Growing Appetite for Poultry." *International Food and Agribusiness Management Review* 18, Special Issue A (2015): 115–36.

Yan, Hairong. "Neoliberal Governmentality and Neohumanism: Organizing *Suzhi*/ Value Flow through Labor Recruitment Networks." *Cultural Anthropology* 18, no. 4 (2003).

Yan, Yunxiang. "Food Safety and Social Risk in Contemporary China." *Journal of Asian Studies* 71, no. 3 (2012): 705–29.

Yan, Yunxiang. *Private Life under Socialism: Love, Intimacy, and Family Change in a Chinese Village, 1949-1999*. Stanford, CA: Stanford University Press, 2003.

Yang, Jie. *Unknotting the Heart: Unemployment and Therapeutic Governance in China.* Ithaca, NY: Cornell University Press, 2005.

Yang, Jingqing. "Professors, Doctors, and Lawyers: The Variable Wealth of the Professional Classes." In *The New Rich in China: Future Rulers, Present Lives*, edited by David Goodman. London: Routledge, 2009.

Yang, Mayfair. *Gifts, Favors, and Banquets: The Art of Social Relationships in China.* Ithaca, NY: Cornell University Press, 1994.

Yuan, Haifeng [袁海峰]. "基层畜牧兽医站管理的一些做法和体会." 兽医动态 12, no. 8 (1986): 55–56.

Zeng, Nanjing, Guanhua Liu, Sibiao Wen, and Feiyun Tu. "New Bird Records and Bird Diversity of Poyang Lake National Nature Reserve, Jiangxi Province, China." *Pakistan Journal of Zoology* 50, no. 4 (2018): 1199–600.

Zhan, Mei. "Civet Cats, Fried Grasshoppers, and David Beckham's Pajamas: Unruly Bodies after SARS." *American Anthropologist* 107, no. 1 (2005): 31–42.

Zhan, Mei. "Wild Consumption: Privatizing Responsibilities in the Time of SARS." In *Privatizing China: Socialism from Afar*, edited by Aihwa Ong and Li Zhang. Ithaca, NY: Cornell University Press, 2008.

Zhang, Bo, Jinchao Qiu, Yufeng Wang, Chengcheng Luo, and Anqiang Xiang. "从'稻田养鸭'到'稻 鸭共生': 民国以来'稻田养鸭'技术的过渡与转型－以广东地区为中心." 农业考古 3 (2015).

Zhang, Letian, and Tianshu Pan. "Surviving the Crisis: Adaptive Wisdom, Coping Mechanisms and Local Responses to Avian Influenza Threats in Haining, China." *Anthropology and Medicine* 15, no. 1 (2008): 19–30.

Zhang, Qian Forrest, and John A. Donaldson. "The Rise of Agrarian Capitalism with Chinese Characteristics: Agricultural Modernization, Agribusiness and Collective Land Rights." *China Journal*, no. 60 (2008): 25–47.

Zheng, Hong'e, Li Xiaoyun, Wang Libin, and Yu Lerong. "对乡村社会风险管理体系及存在问题的反思: 以禽流感的风险应对为例." *Journal of Nanjing Agricultural University* (Social Sciences Edition) 10, no. 4 (2010): 113–20.

Zheng, Yougui [郑有贵] and Li Chenggui [李成贵], eds. 一号文件与中国农村改革. 合肥市: 安徽人民出版社, 2008.

Zou, Xiuqing [邹秀清]. 鄱阳湖农业: 自然资源利用演变机制研究. 南昌: 江西人民出版社, 2008.

Beck, Ulrich, 196
Beveridge, William Ian, 31, 38
biosecurity: overview of, 71; and farm
    closures, 93, 233n86; and farm size,
    76–77; global-local tensions, 161; and
    industrial agriculture, 73; polysemy of,
    229n28; and poultry production sectors,
    72; and vital uncertainty, 94–95.
    See also Food and Agriculture Organ-
    ization biosecurity
biosovereignty. *See* national biosovereignty
bird flu. *See* avian influenza
blame. *See* geography, of blame
Bourdieu, Pierre, 90
Bresalier, Michael, 32, 218n13
bridge species, 116
Burnet, Frank Macfarlane, 40–41, 194, 244n8

Canguilhem, Georges, 51–52, 200, 215n24
Castellan, David, 157, 169
Center for Disease Control, U.S. (CDC), 157,
    160, 161
Centers for Animal Disease Control
    (CADC): name of, 240n7; research
    projects, 140–41; surveillance programs,
    148–49; trader trust of, 141–42; transition
    to, 160–61; and vaccination debates,
    147–49
Centers for Disease Control and Prevention
    (China CDC), 9, 160, 209
chapter overviews, 22–23
Chen Hualan, 128–30, 143, 148
chickens: and agricultural development, 15,
    16f, 65, 72, 77–78, 84, 149, 179, 229n35;
    and flu outbreaks, 4, 20, 56–57, 89,
    179, 191; culling of, 4–5, 70, 191; in lab
    research, 41; as virus host, 4, 6, 50
chijiao shouyi, 164, 187–90
China: the Cultural Revolution, 15, 78; and
    global health regimes, 125–26; labora-
    tory infrastructure, 61; market economy
    shift, 14–15, 65, 75–78, 108, 164–66;
    outbreak nonreporting, 8–9, 125–26;
    and pandemic preparedness, 157–58;
    pandemics originating in, 2; rural issues,
    the three, 18; SARS, 8–9; and the United
    Nations, 34, 43; and the WHO, 34–37,

43–44, 126, 220n39; wildlife trade bans,
    210
China, the Republic of, 7, 34, 43, 220n39
China Vietnam Forum on HPAI Manage-
    ment and Control, 147–48, 151–52
China Wild Animal Protection Committee,
    113–14
Chu Chi-Ming. *See* Zhu Jiming
conservation, 106
context as scientific object, 24, 196
coronaviruses, 209–12, 247n1
COVID-19, 209–11, 247n1
culling, 69–70, 93, 191
cultural difference, 153–54, 197
Cunningham, Andrew, 28

daotian yangya, 46, 50–51, 54, 57, 66–67, 95f
Daszak, Peter, 193
dayan. See swan geese
Deleuze and Guattari, 191, 207, 217n54
Deng Xiaoping, 75
detachment: and field epidemiology, 181;
    and laboratories, 99, 202; official
    veterinarians, 162, 189–90, 197; and
    Poyang Lake research, 103; problems
    with, 195; by professionalization, 162; in
    science, 99
development narratives, 112
diagrams, 67, 227n8
disease ecology, 49, 194, 197, 215n21, 222n62
disease prediction, 204–5, 212
displacement, scientific: overview of, 19,
    114–15; conceptual development of, 20;
    and experimental systems, 19, 201; field,
    100, 103, 115–18; in HPAI research, 57; and
    laboratories, 19–20; and new assem-
    blages, 207; and nonexperts, 21–22; and
    objectification, 52–53; and the pandemic
    epicenter, 196, 201; at Poyang Lake,
    20–21; and research objects, 115, 201; and
    scale, 19–20; and scientific ideology,
    52–53; as socio-spatial process, 21; spatial,
    19, 61; unprecedented events, 114–15
domestic-wild interface: distinction difficul-
    ties, 98, 114–15, 119; pandemics, enabling,
    97; at Poyang Lake, 18–19, 103, 105, 115–17
dongwu weisheng jiandusuo, 158, 176

feed industry, 81
feeding practices and social relations, 81
Fenner, Frank, 222n65
Fidler, David, 8
field epidemiology, 157, 181
Field Epidemiology Training Program
    for Veterinarians (FEPTV): and access
    through affinity, 197; development of,
    157; epidemiology in, 169–70; farm
    visits, 170–71; first module, 158–59,
    177; funding sources, 159; methodol-
    ogy, 169, 177; and national reform, 159,
    181; and official veterinarians, 169, 175;
    organizer concerns, 174–75; outbreak
    investigation, 170–71, 176; as *peixun*
    course, 171–73; as professionalization,
    169; science-state relationships, 175–76;
    statistics methods, 170–71; and *suzhi*,
    169, 172; trainees, 158–59, 172–73, 176–78;
    and the U.S. CDC, 157, 161
Fischer, Michael, 10
Fleck, Ludwick, 200–201
folk vets, 163–64
Food and Agriculture Organization (FAO):
    creation of, 6; field research manual,
    116; international *versus* national staff,
    135; John Boyd Orr on, 133, 239n37; limits
    of, 133; meat production statistics, 15, 16f;
    mission of, 6; pandemic origins defini-
    tion, 6; pandemic prevention strategies,
    67–68; poultry production typology, 72,
    229n36; Poyang Lake project, 138–39;
    relationship building, 139; Technical
    Cooperation Program (TCP), 133–34;
    technical internationalism of, 134–35, 139.
    See also Emergency Center for Trans-
    boundary Animal Diseases (ECTAD)
Food and Agriculture Organization and
    avian influenza: animal health focus,
    6–7; collaboration, calls for, 5; control
    strategies, 6, 67–68, 132; infection sources
    identified, 68, 227n12; involvement justifi-
    cation, 68; One Health strategy, 5; Poyang
    Lake project, 138–39; TCP assistance,
    133–34; vaccination guidance, 228n26
Food and Agriculture Organization bio-
    security: overview of, 69, 71; agricultural

restructuring, 70–74; containment, 70;
    and displacement, 95–96; and economic
    transition, 71–72; eradication, 69–70;
    introduction to, 228n27; local solutions,
    195; poultry sector restructuring, 71–73,
    86; reactions to, 74; vaccines, 128, 131
foot-and-mouth disease, 125, 153–55
forestry expos, 113–14
Fortun, Kim, 216n37

Gell, Alfred, 90
geography, 53–54, 60; of blame, 23, 49, 197
global health: deindustrialization calls,
    205; diplomacy, 126; global suitability,
    204–5; Global Virome Project, 203–4,
    206, 210; network model, 204, 206, 211;
    One Health concept, 5, 49, 102, 213n8;
    One World concept, 5, 213n8; origins
    of, 2; planetary health model, 205–7;
    preparedness programs, 161; security
    approaches, 68, 227n17; and sovereignty,
    9–10, 126; sphere metaphor, 206.
    See also World Health Organization,
    World Influenza Program
Global Initiative to Share All Influenza
    Data, 210–11
globalism: and environment, 202; and infra-
    structure, 28; network or sphere, 204–5
global suitability models, 200–205
Global Virome Project, 203–4, 206, 210
ground-truthing, 61, 151
guanfang shouyi. See veterinarians, official
guanxi, 138–39. *See also* affinity access
    strategies
Guan Yi, 101
guimo chang, 74–77, 248n53. *See also* scale and
    scattered farm types
Guo Yuanji, 45, 61

H5N1. *See* avian influenza
H5N2, 129–30, 191, 199
Harbin Veterinary Research Institute, 128,
    130; post-vaccination surveillance, 146,
    148–49; sample analyses, 141; vaccine
    development, 92, 129–30, 148–49;
    virological sequencing, 142
Harrison, Mark, 27

planetary health, 205–7
poultry censuses, 59, 226n41
poultry production typologies, 72, 229nn35–36
Poyang Lake: agriculture, 14–15, 18–19, 21, 66, 76–82, 106; the FAO and, 138–39; and the livestock revolution, 14; location of, 12, 13f; Mao era waterworks, 12–15, 65; and the market economy shift, 15, 65, 106; Migratory Bird Refuge, 15, 17, 106, 110; migratory birds, 11, 14, 106; pandemic epicenter at, 12, 101, 138, 193; as flu research site, 12, 20–23, 67, 77, 97–98, 102–5, 115–16, 138–39, 151, 225n38; and scientific displacement, 20–21, 69, 100, 116–19, 195, 202; as virus source, 11–12, 67, 101, 192–93. See also avian influenza and Poyang Lake; duck farming
production time, 89–90
Prosser, Diane, 15
proto-ideas, 200
public health, 8–9, 160–61

quality. See suzhi

Rabinow, Paul, 215n22, 243n50
reassortment: definition of, 2, 41; factors encouraging, 43, 97; interspecies, 43; in natural settings, 47; as pandemic source, 42; in vaccine development, 129
recombination, 40–41, 222n65
Redfield, Peter, 127
remote sensing. See satellites
reservoir of viruses: in animals, 2, 6, 38–41, 43, 48, 50, 66, 200, 209, 211, 227n12; in China, 2, 6, 29, 38–40, 43, 53, 66, 200; and geography of blame, 23; and Global Virome Project, 204; for COVID-19 pandemic, 208, 211. See also animals; pandemic epicenter hypothesis; pandemic origins
restructuring, economic, 71–73
reverse genetics, 129–30
Rheinberger, Hans-Jörg: epistemic things concept, 52; on experimental systems, 19; scientific displacement model, 19–20, 114–15

rice-duck coculture, 46, 50–51, 54, 57, 66–67, 95f
rice land use laws, 79–80
risk maps: development of, 54–59; and ecology of influenza, 199–200; Emergency Center use of, 145–51; versus the epicenter hypothesis, 198; global, 199–200; versus point-of-origin theories, 201; satellite use for, 58–60, 150–51, 197–98
Robinson, Timothy, 204–5
rural land categories, 79–80
Russian flu, 27–28

Sanlitun district, Beijing, 7–8, 11, 134, 143, 147, 156
Santos, Gonçalo, 86
sanyang, 74–75, 86, 131, 248n53. See also scale and scattered farm types
sars-CoV-2, 247n1
satellites: and bird tracking, 12, 97, 101, 117; in disease ecology research, 101–2, 104; and ground-truthing, 151; landscape monitoring, 58; rice cropping monitoring, 151; and risk mapping, 58–60, 150–51, 197–98
scale and scattered farm types, 74–76, 78, 86; in vaccine policy, 131, 248n53
Schild, Geoffrey, 42
Schmalzer, Sigrid, 164
Schneider, Laurence, 175
science: anthropology of, 20; as becoming, 203; detachment in, 99; development-by-accumulation, 52, 224n16; epistemic things, 52; in experimental conditions, 215n24; in the field, 99–100, 115; field versus laboratory, 99–100; historiography of, 174–75; knowledge production, 19; local contexts, 99; paradigm shifts, 52, 224n16; relations with state, 175–76; society, relation to, 195; sociology of, 174–75. See also displacement, scientific
scientific change, 52–53
scientific ideologies, 51–53, 200
severe acute respiratory syndrome (SARS), 8–9, 160
Shapin, Steven, 243n50
Shibahu village, 79

vertical integration, 73, 76, 84
veterinarians: and administrative reforms, 165, 167; barefoot, 164, 187–90; common program, 163; Mao era, 160, 163–64; detachment by professionalization, 162; *versus* duck doctors, 190; and epidemiology, 157, 159–60; epidemiology training, 140–41; epistemological boundary crossing, 206–7; folk vets, 163–64; at Hong Kong influenza conference, 45; informal, 162; laboratory expertise subordinating, 21; and the livestock revolution, 156–57, 165; and medical doctors, 9, 137; and pandemic preparedness, 158; poultry industry perspectives, 74–76; professional, 167; stratification of, 206–7; and surveillance, 55; *suzhi* discourses, 168; vaccination work, 131; value to experts, 22. *See also* Animal Husbandry and Veterinary Stations; duck doctors; Field Epidemiology Training Program for Veterinarians
veterinarians, official: overview of, 167–68; and avian influenza response, 161, 167; creation of, 160, 167; criticisms of, 175, 178; detachment of, 162, 189–90, 197; *versus* duck doctors, 190; and farmers, 207; field epidemiology training, 169, 175; and specialist knowledge, 162; and veterinary reform, 167–68
veterinary administration, 163. *See also* Animal Husbandry and Veterinary Stations
Veterinary Bulletin, 146
Veterinary Bureau, 74–76, 147, 167, 238n23
veterinary reform: overview of, 160–61; and avian influenza, 160, 167–68; county-level, 167; critiques of, 166, 180–81; employee reduction, 181–82; global preparedness, 161; and global preparedness programs, 161; *guanjia* to *guanfang* system, 166–67; and market economy transition, 166–67; MOA bureau, 167; and professional stratification, 162–63, 180–81; *suzhi* discourses in, 168–69, 182; Veterinary Bureau creation, 167
veterinary science teaching, 143–44
Vietnam, 73, 229n37

viruses: filter-passing, 29, 218n13; genetics research, 40–41
virulent zones. *See* zones of virulence
vital uncertainty: overview of, 69, 87; in agriculture, 89; and biosecurity, 94–95; disease, 87–91; failed farms, 87–88; and free grazing, 94–95; and insurance, 93–94, 233n93; labor *versus* magic model, 90; of market ducks, 87; and pharmaceuticals, 90–91; and production time, 89–90, 94; and vaccination, 91–92; and wealth, 87, 89

Wald, Priscilla, 197
Webster, Robert: on the 1997 pandemic, 4; Beijing conference proposal, 44–46; career of, 41; in China, 44; hybrid virus discovery, 41; on pandemic sources, 42, 48; reassortment discoveries, 43, 47; virus reservoir findings, 48
wild animal farming, 107–8, 119–20, 209–12
wild animals and COVID-19, 209–12
wild birds: deaths of, 100–101; farmed, 116–18; transmission confirmation, 101; vector or victim question, 100. *See also* migratory birds; swan geese
wild food, 107
wildlife markets, 209–10, 212
wildlife trade, 210–12
Woods, Abigail, 70
working landscapes: 14; and scientific displacement, 22–23, 69, 98, 100, 118–20, 195; veterinary detachment from, 162, 168–69, 181, 189–90
World Health Organization (WHO): and avian influenza, 5–6; and China, 34–37, 43–44, 126, 220n39; and COVID-19, 209; SARS recommendations, 8; Veterinary Public Health division, 39; and virus sample sharing, 9
World Influenza Programme (WIP): and the 1957 pandemic, 33–34, 36, 38–39; animal research, 40; goals of, 31–32; international network of, 32–33; origins of, 2, 31; passive surveillance, 55; planetary perspectives, 33; shortcomings, 34
World Organization for Animal Health (OIE), 4–5

Wucheng (Jiangxi), 12, 14–15, 116, 119
Wuhan, 209, 211

xianchang, 240n2
Xiaolan egg market, 78, 82–83, 91
xumu shouyi zhan. See Animal Husbandry and
    Veterinary Stations

yewei. See wild food
Yiwu, 113

zhuanyehu, 78–79, 165
Zhu Jiming, 35–38, 44–45
zones of virulence, 10, 215n22
Zuoyue Wang, 174–75